The Cooperation Phenomenon

Prospects for small firms and the small economies

Dedicated to
Paula, Niall and Maeve

The Cooperation Phenomenon

Prospects for small firms
and the small economies

Edited by Dermot O' Doherty, EOLAS
(The Irish Science and Technology Agency)

Based on a European Conference held in Dublin,
Ireland, 6-7 November 1989

Graham & Trotman Limited
Sterling House
66 Wilton Road
London SW1V 1DE
UK

Kluwer Academic Publishers Group
101 Philip Drive
Assinippi Park
Norwell MA 02061
USA

ISBN 1 85333 396 4

First published 1990

British Library Cataloguing in Publication Data

O' Doherty, D. (Dermot)
The cooperation phenomenon : prospects for small firms and
the small economies
1. Cooperatives
1. Title
334

ISBN 1-85333-396-4

Library of Congress cataloguing-in publication data is available from the publisher

Printed and bound in Great Britain by
Billing & Sons Ltd, Worcester

Preface

This book aims to reinforce and enhance the practice of cooperation for innovation which it addresses at both the analytical and policy levels. Cooperation is particularly important for small firms and small economies whether in terms of formal partnerships and informal networks between firms of equal size or those between small and large firms – both nationally and internationally. These small players need the complementary assets and resources that cooperation brings if they are to become and remain competitive in the current technological and economic climate.

This book shows that small firms and small countries can bring unique characteristics and advantages to the negotiating table and that the public authorities can materially assist and encourage cooperation at the regional, national and international levels.

I believe the ideas and proposals presented here will stimulate discussion and debate on company and national strategies as well as on the role of the European Commission and the OECD, for example, in both developing our understanding of and designing appropriate policies for SME cooperation. It is important at this juncture in the development of cooperation as a global phenomenon that such a discussion should emanate from a small European country with an industrial structure which is overwhelmingly composed of small firms.

J. M. McBride
Chief Executive
EOLAS
Dublin
July 1990

Acknowledgements

I would like to thank the contributors to this book and the speakers at the conference of the same title in Dublin in November 1989 on which it is based, as well as the sponsors (the European Commission) and patrons (OECD) of that conference. I would also like to express my gratitude to the members of the conference committee and the rapporteurs, particularly Graham Vickery, OECD and Jurgen Allesch, TVA Berlin, who acted as both but were not actually speakers at the conference itself. I am grateful to Renée Browne and the members of EOLAS Industrial Education Programme for their unstinting efforts to make the conference a success and to my colleagues in the Science and Technology Policy Group at EOLAS for their help and support. Paul Holden and Ann Phelan of Rédacteurs, Dublin looked after the design and preparation of the book with their usual professionalism. Finally I would like to thank my wife Paula for her support in this and all other cooperative ventures.

D.O'D.
July 1990

Contents

Introduction: Small Firms, Small Economies and Cooperation

This book brings together three areas of distinct analytical and major policy concern:

- small firms, where there is a vast literature on their genesis, growth and general role in economic, technological and industrial life;
- small countries and regional economies in which there is renewed interest and an increasing emphasis on various forms of innovative approaches to indigenous economic and industrial developments;
- the growing phenomenon of cooperative, reciprocal relationships between firms and their impacts on economic and industrial competitiveness. While the literature on these issues is growing there is still a relative dearth of analysis on small firm cooperation.

The general approach of the book is based on the hypotheses that small firms have specific needs and characteristics when it comes to cooperating either with other small firms or with larger partners; that small firms and small countries have both advantages and disadvantages in facing the challenges of technological change and the globalization of markets; and that these challenges can be addressed through cooperative mechanisms. These include formal contractual links, often with larger companies (and countries?), and the informal sharing of ideas and information, often with other small firms (and countries?) in the same business as themselves as well as with customers and suppliers who are playing an increasingly important role in the whole process of innovation and technological change. The philosophical position adopted here also assumes that these issues and mechanisms can be addressed and enhanced through the actions of the public authorities at regional, national and international levels e.g., through the European Community (EC) and the Organization for Economic Cooperation and Development (OECD).

In some senses the bringing together of these three analytical and policy fields involves challenging what was, until relatively recently at least[1], the received economic wisdom about all of them. Take small firms - they were seen to be competing in an atomistic way in free markets, with no links between them and with little choice as to whether and how they developed particular forms of strategy and behaviour to deal with

1 Alfred Marshall however refers to the importance of cooperation between firms in his famous *Principles of Economics* 1906 (see Mariti P. in this volume)

their situation. Although the process of changing these perspectives started some time ago[2] the same kind of logic is still often applied to small countries and economies, Here the dominant model is the "small open economy" which has to act as a "price-taker" in international markets. The small countries have been seen as having little or no control over their level of "given" comparative advantage at macro or micro level. This is now changing with the shift towards the concept of "competitive" advantage where the intangible assets of technological, design and marketing capabilities are coming to the fore as the main factors defining the level of international competitiveness[3].

Equally, different forms of cooperation and collaboration, or even "collusion" in the language of the anti-trust expert, were seen to be characteristic of oligopolised markets dominated by a few large firms. It is increasingly clear both from the material presented in this book and from research reported elsewhere[4] that cooperation is very much a "small firm" activity. If we are to agree with Porter[5] that a new definition of firm and national competitiveness is important then in the longer term cooperation may be seen to be of significantly greater importance for small rather than for large firms. The latter could easily find that cooperation may be a limited and limiting strategy whereas for small firms it can provide the complementary skills and resources which are essential for competitiveness and survival in world markets. The reality is that small and medium-sized enterprises (SMEs) do not themselves generally have the range of capabilities and resources to research, commercialise and market new products. They need cooperation with large firms and other SMEs in order to remain in the innovation game.

We must remember that, with the possible exception of new technology-based firms (NTBFs) small firms are normally technology "followers" or imitators. They need formal links with larger companies, e.g. in the form of subcontracting arrangements, to ensure that they receive the most up-to-date technology in a form in which they can use and apply it immediately and with little or no adaption.

2 See references in Dickson, Lawton-Smith and Smith in this volume

3 See Porter, Michael E., "The Competitive Advantage of Nations" HBR March – April 1990

4 e.g. Håkannson H. *Corporate Technological Behaviour – Cooperation and Networks* Routledge 1989

5 See 3 above

Networking between small firms themselves, and personal networks between the technical and marketing members of their staff, are now seen as playing a key role in product development and the opening up of new markets.

With the rapid advances in technology and major changes in markets and in overall competitiveness now characteristic of all European economies there has been a corresponding increase in the effort to devise appropriate policies for small firms in particular. This counter trend to increasingly negative attitudes and approaches to "the state in business" in the 1980s has been reflected at European Community level, as well as in national capitals in a range of technological and industrial measures aimed at enhancing the process of SME cooperation.

These have tended to grow in an ad hoc manner without any integration into an overall industrial or "small firms" policy, but with many of the pieces for such policies being put in place. Consequently, there are now a whole range of measures at national and Community levels for which the whole is *less* than the parts. One of the aims of this book, as it was of the Conference[6] on which it is largely based, is to try to cohere this complex of measures into a truly integrated and comprehensive approach to SMEs in the first instance, and to cooperation as a strategy and approach for SMEs and to look at these in the context of small countries and economies in particular. Specifically the Conference (and similarly this book) had the following aims and objectives:

- To critically examine the recent experience of small firms in networking and cooperation – particularly in the smaller economy context – and the regional, national, and international (including European Community) policies and programmes aimed at encouraging and enhancing that cooperation.
- To facilitate the exchange of ideas and experiences between international, national and regional decision-makers, advisers and researchers who are attempting to understand interfirm cooperation.
- To examine the extent of small firm involvement in such cooperation within and between economies; to assess the costs and benefits

6 "The Cooperation Phenomenon – Prospects for Small Firms and the Smaller Economies" International Conference organized by EOLAS, sponsored by the European Commission and held under the patronage of the OECD: Dublin, November 1989.

involved and to identify constraints which are hampering the further development of networking and cooperation by small firms and the smaller economies.

- To define the institutional mechanisms and negotiating positions necessary to maximise small firms involvement and the share of the benefits accruing to small firms from interfirm cooperation.
- To make specific proposals for policy, based on the strategic options facing small firms and the smaller economies.

European-level policies were a major focus in the context of the need for novel approaches to greater "cohesion" or equality of development between the regions and states of the Community.

The Conference was attended by 100 international, national and regional researchers, advisers and decision-makers, including industry representatives, who were interested in analyzing and maximising the benefits of networking and cooperation for small firms. While no consensus emerged from the discussions on either the diagnostic or the remedial side many key issues and policies were aired in an open and forward-looking way. In the absence of either an agreed analysis or clear solutions people were reluctant to advocate radical policy approaches. At the same time they clearly recognised that the problems for SMEs in carrying through cooperation strategies were serious and deep-seated and that existing policies were not adequate in their scope and impacts. This book goes somewhat further than the conclusions of the Conference itself in outlining a research and policy agenda for the 1990s addressing the small firms; small economies; cooperation complex. Like any other agenda it can and probably should be altered but the need for an agenda on such an important set of topics can hardly be doubted.

Section 1: Small Firms and Cooperation

Cooperation manifests itself in a variety of approaches and contexts. Interfirm cooperation, which is the major focus of this book, covers a spectrum of agreements and arrangements from formally negotiated contracts between two firms to the exchange of knowledge and knowhow within informal industrial networks. This diversity is reflected in the literature on interfirm cooperation and a number of definitions of "interfirm cooperation" drawn from this literature are given in Appendix 1 to this book.

The focus in this section is not just on interfirm cooperation in general but on how small and medium sized enterprises (SMEs) participate in that process. Small firms differ from large firms in the way they approach innovation in general, of which cooperation can be seen as a subset (see Appendix 2 to this book). It should also be remembered that there are an infinite variety of SME types. The categorization that is most relevant here is that into new technology based firm (NTBFs) and SMEs in general. Given their relative weakness in marketing and financial terms NTBFs are often willing to enter "complementary" relationships to compensate for these weaknesses. They also tend to have something to offer when it comes to informal exchanges.

Naturally the picture with regard to SMEs in general is much more mixed with those with a customised product or some unique capability having something to bring to the cooperation party or needing a partner to carry out the service element required for the "delivery" of the product to their often distant customers.

Both kinds of small firm indicate a preference for entering into formal "horizontal" partnerships until other small firms but in practice many formal agreements tend to be with larger firms, for example to gain access to wider markets and to get the fullest possible return on their technology.

Informal networks between small firms in the same industry are much more common, as are a range of cooperating links with suppliers and customers. User-producer links are becoming a major focus of both innovation and cooperation research (see Lundvall and Dalum in this volume).

The contributions to this section are intended to analyse and illustrate this spectrum of small firm cooperation possibilities and realities – both formal and informal, technological and developmental in nature.

The section starts with case histories of three Irish NTBFs chosen to give an integrated "hands on " view of the typical small company and the state of development of cooperation consciousness and activity in Ireland. These small firms operate very much in a European and international context, with a range of cooperation links with both small and large companies.

It then goes on to give a range of strategic, analytical and operational perspectives on small firm to large firm and small firm to small firm relationships from an international management consulting group and academic researchers from Italy and Britain respectively. The first paper highlights the mutual advantages for the large firm and SME in cooperating, outlines the role of technology in company development and stresses the importance of sales and marketing agreements and related public policy initiatives in addition to technology programmes more narrowly defined. Drawing on some empirical evidence which is detailed in the second half of the paper Mariti's contribution offers a densely-argued theoretical perspective on the nature of interfirm cooperation, particularly between small firms. The British paper offers a blend of survey and analytical material outlining the management motivations for and against involvement in interfirm cooperation.

Two of the papers in this section stress the importance of intermediaries in the cooperation and technology transfer process. The next contribution gives the results of a recent study of technology transfer intermediaries in France, and indicates a growing role. The two final papers analyse "the licensing experience" in Portugal and Ireland. In both countries the level of formal agreements is surprisingly low and particularly in Portugal the evolution of the Single Internal Market could even reduce this level. Lack of experience in negotiating agreements appears to be a constraint in the Irish context.

Mentec International Limited – A Case History

M.A. Peirce

Mentec is an Irish owned computer systems and computer products company. It was founded in 1978 by the current chief executive, with the objective of addressing the industrial marketplace and its needs for computer-based systems and products.

Initially targeted at manufacturing applications, it broadened its activities to address the full spectrum of computing requirements of manufacturing companies and now addresses financial, accounting, distribution, manufacturing and control systems computing.

Mentec activities are organized within a Group which embraces five activities under 5 group companies.

1. Mentec International Ltd – A Dublin-based systems company.
2. Mentec International (UK) Ltd – the UK equivalent of (1), which is based at Warrington, near Manchester.
3. Mentec Computer Systems – A Dublin-based company which designs and manufactures computer products, processors, data acquisition equipment and networking products.
4. Mentec Sales Ltd – supplies add-on equipment, products and systems to customers in Ireland and overseas.
5. Computerent plc – A majority-owned subsidiary, based in London which leases and rents computer systems.

Summary Profile

Founded in 1978 under the IDA's Enterprise Development Programme

- Located in Dublin, Manchester and London
- 150 employees: 105 in Ireland
- Mentec have 20 overseas distributors, principally in Europe but also in Japan and Australia
- Sales revenues of £14m in 1988-89
- Anticipated sales revenues of £18-£20m in 1989-90
- 1989-90 spend on R & D £700,000
- 70% of sales are to overseas markets

Our client base has always tended to be the larger manufacturing enterprises, utilities and the manufacturing wings of major international companies. Starting up our company in Ireland, this was an obvious target market with the influx of the IDA-promoted companies which has been maintained over the past twenty years.

Additionally, we targeted on the larger indigenous industries and achieved substantial success in the food and beverage sectors. Our UK company has maintained this approach and now works with some of the world's largest companies.

Mentec works very closely with Digital Equipment Corporation. In parallel with the Mentec emphasis on Digital brand names, Mentec has assembled a portfolio of branded applications software products, software and development tools.

This portfolio includes computer industry brand names such as CODA, Impcon, LEX, 20/20, Powerhouse, MIDAC, PHS, Mensa and Network Professor. These products are marketed internationally by Mentec and many other companies in the computer industry. MENSA and MIDAC are Mentec proprietary products. The Network Professor product is the result of an inward technology transfer from the U.S. with Mentec having exclusive manufacturing and marketing rights in Europe. The remaining brand names are products which we sell and support under licence.

Link with Digital

Our association with Digital started very early in the life of the Company, in fact one year after our start up. We had been working with two other companies, one a minicomputer manufacturer and the second a manufacturer of processors and electronic components.

The search to find a single architecture around which to build our commercial and technical systems led us to Digital. We made a simple judgement that Digital had placed a massive emphasis on Operating Systems development and that this would guarantee them a place amongst the winners.

This move to Digital technology required substantial investment in equipment and training. Our relationship with Digital commenced with conventional "third-party" Agreements, referred to in the industry as OEM Agreements. In my view it has developed into a much closer relationship than is legislated by such Agreements. For example, we conduct a considerable amount of joint marketing, joint quotations, joint sales and joint involvement in subsequent project execution. We promote joint seminars and discuss future business trends. The Digital Ireland staff have been exceptionally promotive of our endeavours to sell overseas. In exceptional instances, when we have been dealing with very large prospective customers who would have had anxiety over our size, Digital have underwritten our performance in advance.

In the period since our link with Digital started, Digital have been outstandingly successful worldwide, in Europe and in Ireland. Without question, our ability in Mentec to work and trade internationally has been massively assisted by the fact that Digital, PDP and VAX are household names throughout the computing world. Mentec is able to provide systems to a multinational client in America, France and Italy based on an original system shipped to their UK facilities, simply because the VAX architecture has been adopted by that client as their application building block.

The scale of R&D conducted by Digital illustrates the background activities on which Mentec levers in the provision of DEC-based systems.

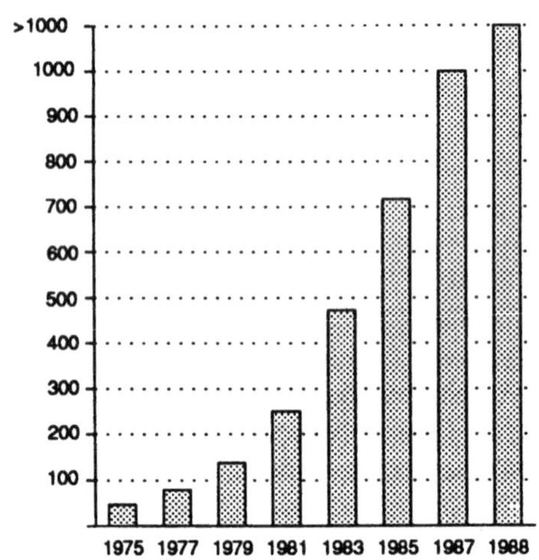

Chart 1: R&D Expenditure by Digital 1975 - 1989

An estimate is that over the 14-year period 1975-1989, Digital will have spent some $5.75 bn on R & D, resulting in either hardware or software sold under the PDP and VAX brand names – as processors or multicomponent systems.

Further Partnerships and Relationships

These fall within two broad categories:

a) Commercial and trading frameworks.

b) Research and Development frameworks.

In the first category, I would include the Distributor and Technology Transfer partners to which I have already referred.

The second area includes interactions within the EEC Esprit and Brite programmes, together with relationships which promote and assist our R & D programmes, including links with academic institutions.

Generics Software Limited – A Case History

R. Cochran

Background

Generics was established in 1984 in order to address a very specialized high technology niche of the computer software market. It set out to address an international marketplace from its inception.

The company's mode of operation is to undertake:

- R&D projects (both publicly funded and commercial);
- consultancy and training in its area of specialization;
- development of new innovative products to at least advanced prototype stage; which can then be licensed on to suitable companies for full commercial exploitation.

The company's area of specialization is in the area of Software Engineering (i.e. the process of software development rather than the end-user application of software). In particular the company aims to bring the most up-to-date research advances (where relevant) to bear on current practice in order to ensure better control, quality and productivity for the software development process.

Initial Strategy

Generics was founded by 2 people – one of whom has a background in a variety of consultancy/user positions while the other was an academic expert in the proposed area of specialization. Thus from the start it was natural to maintain a close link with the university. This involved recruitment of graduates, using the college as a sub-contractor, maintaining a physical telecommunications link between the two computers, etc. All of this helped to create the high-technology culture which was required.

The second aspect of this initial strategy was the fact that Generics effectively launched the company by participation in three EC-sponsored projects. This not only assisted in the necessary working capital requirements of the early stages but also meant that the company made contact and developed relationships with relevant people in France, Belgium, Germany and other member states, and thus quickly established a network of relationships internationally.

Finally, it should be mentioned that the company maintained and extended participation in international technical fora by its staff, thus widening this international network beyond that possible otherwise.

In summary, the company was successful in building up from an early stage a variety of Europeanwide networks and partnerships. These served both to assist in the primary objectives of the company but also in themselves to establish a "profile" for the company across Europe.

Evolution

As the company evolved, other forms of cooperation also became necessary. In particular, more commercial rather than technical cooperation mechanisms were needed.

The company made the usual contracts to try to establish distributors in the main European markets. The nature of our output, however, did not make this very satisfactory. We therefore turned to a more flexible "agency" type of arrangement whereby a company in such markets would represent our interests and seek to develop commercial opportunities for us. This worked well in Germany and Netherlands, where we found small companies, similar to our own but with complementary skills and profiles.

These two bilateral arrangements in turn then formed the nucleus of a wider commercial network of about 7 companies aiming to address the full European market by appropriate collaborative working. As this group is currently just "finding its feet" it is too early to say whether or not it will be successful.

Future Developments

Our experience to date has indicated that, since our market is international and very specialised, we are unlikely to be able to handle our marketing effectively from Ireland. This is particularly so since Generics' product is primarily intangible i.e. technology.

As such, the company has taken the view that it should seek a "strategic partner" – likely to be a large corporation – which can better address the marketing of our technology and probably also invest in its future evolution. At present a number of potential candidates for this role have been identified and have expressed interest. Time will tell whether this is satisfactorily negotiated.

Conclusions

This paper has attempted to show how Generics has used a variety of cooperation mechanisms as it has evolved in an attempt to maximize its position in its chosen market.

Baltimore Technologies Limited - A Case History

J. Mountjoy

The Company

Baltimore Technologies Ltd is a software house which specializes in developing technical software for technical customers to their specifications. The company, founded as a one man operation in 1976, currently employs 28 graduates of engineering and computer science. We expect to have sales in 1989 in excess of IR£1m, all of which is added-value.

Baltimores' customers are product and/or system developers in the datacommunications field and the service provided by Baltimore to them is essentially expertise. We undertake for our customers or participate with them in their product and/or system development activity. During 1989, we will have undertaken this type of work for customers in five European countries outside of Ireland.

Consultancy, in the sense of undertaking studies for customers and producing reports for them, is another activity undertaken by Baltimore. It is not an activity that we have concentrated effort on promoting to date but, in the light of persistent strong demands for our expertise, we are reviewing our approach to it.

Linkage Arrangements

Collaboration, partnerships and a range of linkage arrangements with other companies lies at the core of the business we do. In undertaking development contracts for customers it is not at all unusual that we would accommodate some of the customers' personnel on our development teams and assume project management responsibility for them. Another example is in the area of making joint bids with companies. Baltimore is a totally independent software house which develops a very wide range of equipment, as specified by the customer. In competitive bidding situations, however, particularly for system solutions, Baltimore will now always align itself with a hardware supplier for the purposes of the bid.

Two further examples of the kinds of linkages which are central to the way Baltimore does its business are relevant. It has become apparent to us that, for reasons of time and confidence largely and to an extent for price considerations, our customers favour our using software which has been developed and used previously, whenever possible in providing solutions to them. We can achieve this by acquiring, under licensing arrangements, relevant pieces of software from third parties with which we become proficient and which we re-use as often as possible as kernels of our solutions. These third parties are partners of ours in a very real sense in such situations.

The final illustration has relevance to EC programmes and relates primarily to facilitating the introduction of new technology and to the promotion of standards in so doing. Under the EC's TEDIS programme, Baltimore has been awarded a contract to assist a company called Cargo Community Systems (CCS) with the introduction of an EDI service to a group of users. In this project, CCS is the service provider, the cargo agents and the airlines are the users and Baltimore is a consultant cum catalyst providing assistance to users, in the form of information and advice, and promoting with the service provider and the users the concepts of using OSI and EDIFACT standards.

Role of EC

The European Commission directly and indirectly has to date played an important role in enabling Baltimore to develop the relationships which underpin our business. At the simplest level, we have regularly participated on committees, review groups, advisory groups etc. convened by the EC in areas of relevance to us. The relationships which we have developed with other attendees, who obviously have common interests with us, have proved to be some of the most important we have today. Participation in EC programmes by us has been limited. Our primary focus within the EC is DGXIII and the relevance of their programmes in the past to a company such as Baltimore has been limited. However, having gone through a period of developing the basic technologies we envisage a shift in the EC programmes to application of the technologies and we believe they will become more relevant and attractive to us.

Future

Baltimore is a company which is on a growth path. We have grown quite significantly over the past few years doing the type of things which I have outlined. We are confident in our ability to win and to deliver business in this area and we intend to organize ourselves to grow even more rapidly in the immediate future doing these things. Essentially this means, we have decided on establishing Baltimore companies in a number of countries to expand our market share in those countries. It is distinctly possible that the more attractive and sensible way for us to rapidly achieve this is by entering into joint ventures.

We believe that we can also expand the company by extending its activities into new but related areas. In this context we are currently considering an expanded definition of a consultancy activity, facilities management and software development for a different market. Some of these possible activities would achieve a credible level of business more rapidly if they were to be undertaken as joint ventures.

The Third Channel for Innovation – How Established Companies and Young High Technology Firms Complement Each Other

T. Sommerlatte

The Cooperation Phenomenon is the economic effect of "one plus one equals three". Companies need increasingly to enter networks of alliances and cooperative arrangements ranging from R&D joint ventures to develop new products and processes, to technology licensing and transfer, to market exploration and marketing arrangements.

One of the most effective forms of cooperation is that between

- established companies with a well-functioning sales organization and a broad customer base, but in search of new products and services to ensure growth

  and

- young innovative firms which have developed new technologies, new products and/or new services, but which lack the sales channels to rapidly penetrate the market.

These two types of companies tend to exist next to each other in most of the European economies, not realizing the enormous economic advantages they could gain from cooperating instead of competing with each other.

In the USA and Japan, this kind of cooperation is a key factor in economic growth and innovation.

In fact, it is called, and consciously exploited as, the third innovation channel.

There are essentially three channels for innovation in an economy (see Figure 1).

- Innovation channel 1

  Established companies replace aging products or reactivate stagnant markets through the development of new technologies and products. In order to do so, they have to overcome internal barriers in the form of established but outdated qualifications, habits and structures – they have to nurture innovative developments while maintaining their existing business.

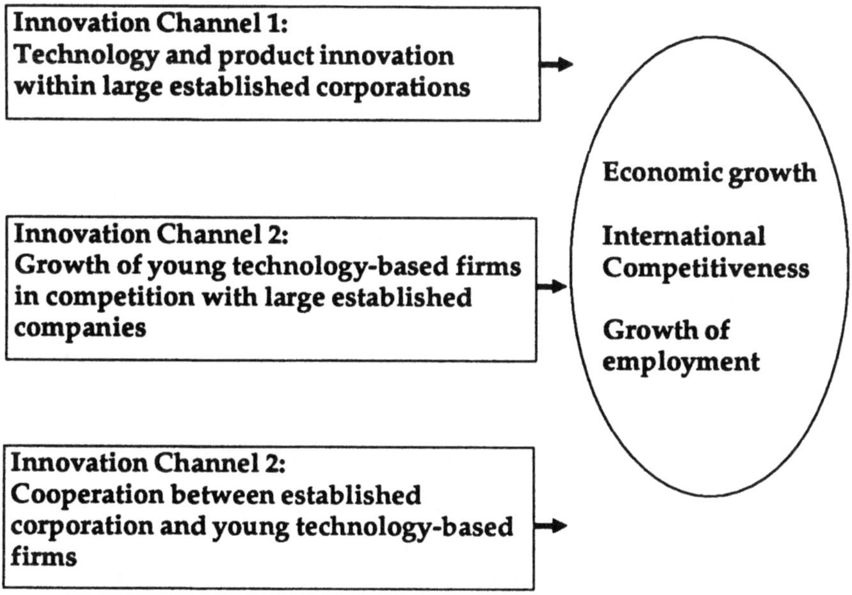

Figure 1: There are 3 channels for innovation in an economy.

We know from experience in technology and innovation management that established companies tend to stick to base technologies and to over invest in them while neglecting or even fighting the pacing technologies of their industry. The reason for this behaviour is that the base technologies used to be key technologies on which these companies built their success in the past and that their culture tends to be averse to technological risk.
As a result, innovation channel 1 is a slow and costly one.

- Innovation channel 2

Young companies are frequently based on new technological solutions and/or innovative thrust looking for market opportunities by substituting and dislodging established products.
In addition to their technological competence, these companies have to develop marketing and sales capabilities, have to under-

stand the market and the need to invest in building a customer base.
More often than not, they underestimate the cost of getting their technological solutions to the market.
Therefore innovation channel 2 is also slow and uncertain.

- Innovation channel 3

  If established and young technology-based companies cooperate, be it in a supplier - customer relationship, be it through cooperative arrangements, then there is a good chance that the technological solutions are channelled through the marketing and sales organization of the established company and reach an existing customer base in an effective way.
  Innovation channel 3 can therefore be a very powerful one.

It is intolerable, especially in times of rapid technological change as today, that innovation channel 3 is used too little in Europe. In fact if economic growth and the growth of employment in Europe have been inferior to that of the USA and of Japan (see Figure 2), then this has to a large extent been the result of a lack of innovation and the high cost of bringing about innovation in Europe.

On average, Japanese companies are 50% faster in getting new products to the market than their European competitors. One of the reasons for this is that they use cooperative arrangements with smaller firms for at least 30% of their technology development requirements.

We believe that Europe can only maintain a competitive position in the major growth sectors (which are those with high technology content and a high level of innovation e.g. microelectronics, information processing, factory automation, speciality chemicals, process-3controlled industries), if we recognize four facts of economic life (see Figure 3):

- Global competition is increasingly based on innovation and the European economies can only prosper if they activate all innovation channels.
- Established companies, even the largest and most competitive ones, cannot maintain technological leadership in all the technology and know-how areas which are relevant for their products.
- Small firms are more effective and inventive in certain technology areas and can realize technological solutions more rapidly, but they do not have the muscle to penetrate markets and to overcome acceptance barriers on their own.

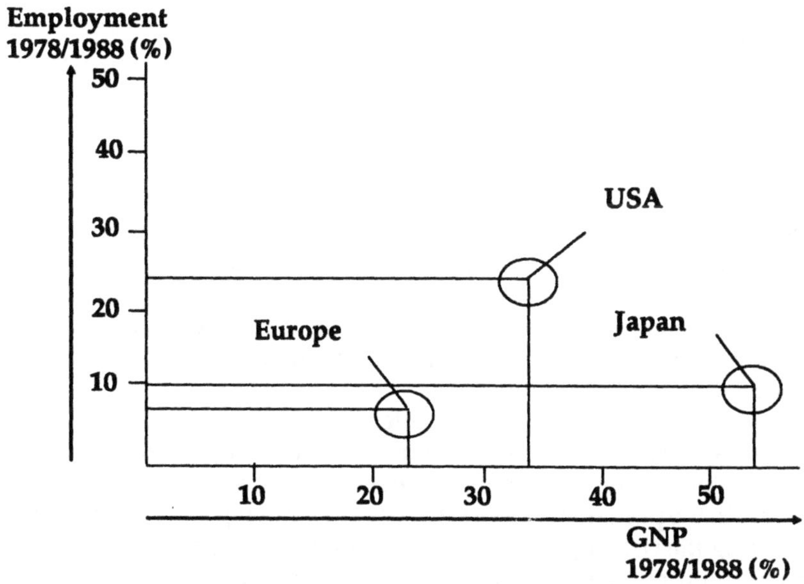

Figure 2: Economic & employment growth in US, Japan & Europe

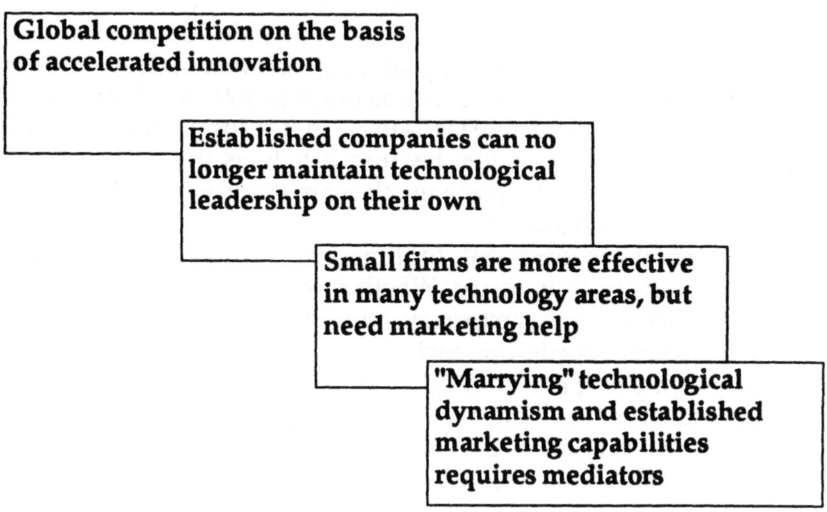

Figure 3: Europe needs to adjust to new conditions of economic dynamics

- By "marrying" the technological dynamism of young entrepreneurial technology firms with the marketing capabilities of established companies innovation processes can be speeded up and economic growth can be enhanced.

Global Competition and Innovation

In most industry sectors we can observe a shortening of product life cycles while the R&D investment per product generation is going up (see Figure 4). This is because many of the major markets have matured, and growth can only be achieved by increasing investment in ever new technologies and products. In this environment, competition is becoming more and more international.

Companies and national economies which are able to produce a continuous stream of innovative and differentiated products and services gain substantial advantage for several reasons.

While in the past, there was a lot to be said for being a "follower" (such as avoiding the risk of pioneering and missing the appropriate product configuration or demand profile), a follower nowadays is finding a much shorter market life time for his products at rapidly sinking prices and is therefore no longer able to recover his development cost, even if he invested less (see Figure 5).

Investigations of Arthur D. Little have shown that exceeding the development time for a new product generation can mean a loss of profits over the lifetime of that product generation of 25 to 30% while exceeding the development cost by as much as 50% in order to meet the target introduction time only leads to a 5 to 10% loss of profitability (see Figure 6). This finding is key because it tells us that companies and national economies must place much more emphasis on the timing of their innovation endeavours.

However, the requirements for timely innovation can be very different from industry to industry. Whether we are talking about satellite systems, office systems or expert systems – the most effective innovation channel is a very specific one each time.

Large complex projects call for sophisticated project management which only large organizations are able to provide. This is true in the aero-space industry, in the automotive industry, in public telecommunication systems and the like. But these large projects normally require a whole series of smaller innovations which the large organizations typically bring about very slowly and at high cost. They therefore need substantial contributions from many small firms. Hence, it is essential

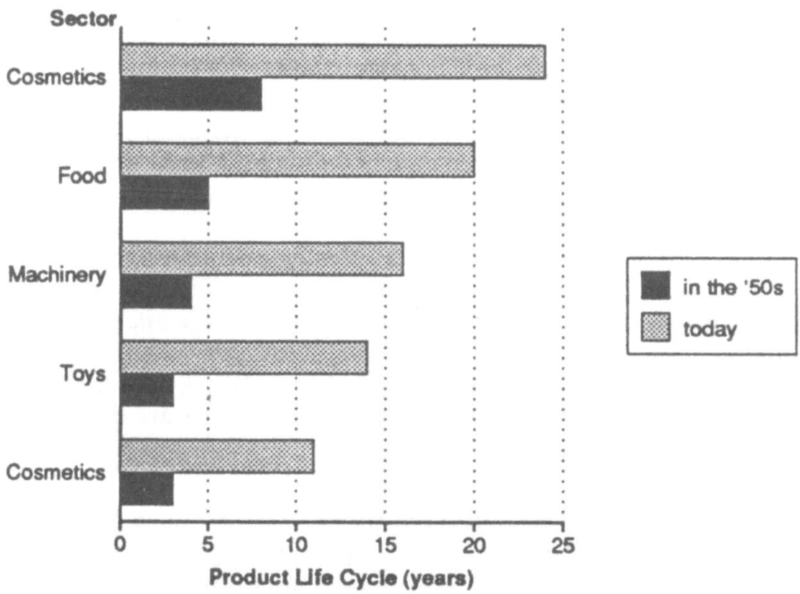

Figure 4: In most industry sectors, product life cycles are shortening

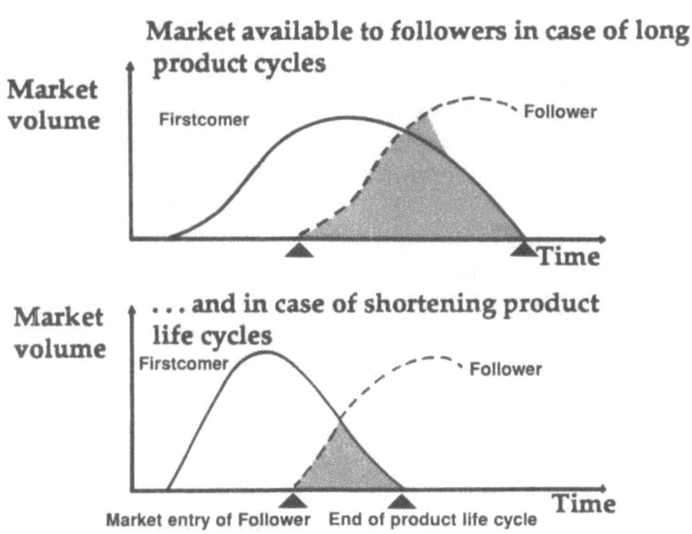

Figure 5: Followers have to live with shrinking market volume and decreasing prices

for timely innovation to effectively organize the interaction of large and small firms as it is done in the U.S. space industry, for example (and much less so in Europe). Strategic alliances of this kind have also been the secret of success in biotechnology. Silicon Valley is a prime example of how the closeness and intensive interchange of young high-tech companies and large electronics corporations can lead to speed and creativity in a key sector, microelectronics.

In Europe we still have to learn. Typically, established companies do not trust young innovative firms. And young high-tech firms fear that they are being exploited when cooperating with large established companies.

This mistrust should no longer have a foundation, because alliances of large established and small innovative companies are clearly in the interest of both sides.

Technological Leadership Has to Be Shared

A growing number of leading companies (IBM, 3M, General Motors, Siemens, ICI for example, to name a few) are finding that they cannot cover all the technology areas where they need to stay at the leading edge of their industry. They have to concentrate their resources on selected technology areas which are key to their business. But there are many key and pacing technologies that have or can acquire strategic importance for them, which they have to monitor and to which they must ensure that they can gain rapid access to them. Take for example a manufacturer of information and telecommunication systems: he will continuously have to watch a wave of technologies, some of them alternative solutions, in order to be sure to offer competitive products and systems at any time (see Figure 7).

The only way to do this is through strategic alliances and partnerships with an array of specialized high-tech companies. The same is increasingly true in the pharmaceutical industry, in the machine tool and factory automation industry, in the car industry and others.

3M has organized a systematic worldwide search for such partnerships and is deriving many of its innovative products from them. Many of our established European companies are still far from getting organized in this respect. They are almost unaware of and inaccessible to small firms and they therefore miss many opportunities of access to new technological solutions. It is significant that leading high-tech firms in Europe rarely experience the rapid growth and financial success that

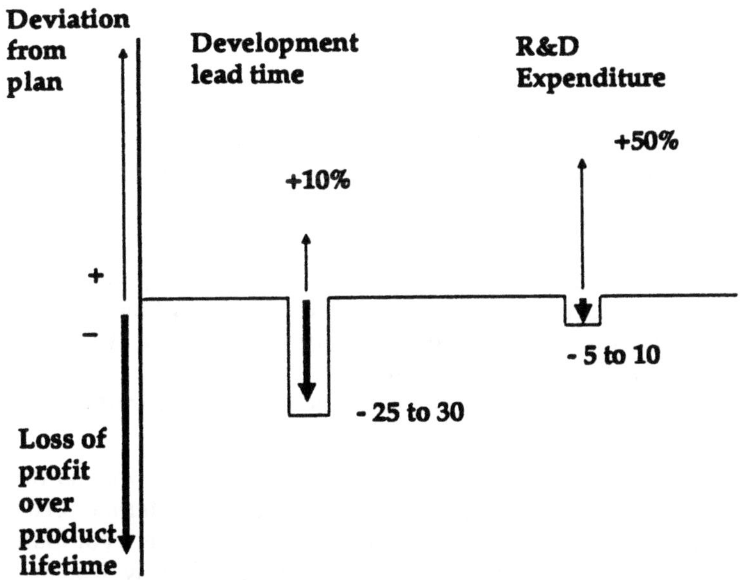

Figure 6: Exceeding development lead times is more costly than exceeding development costs

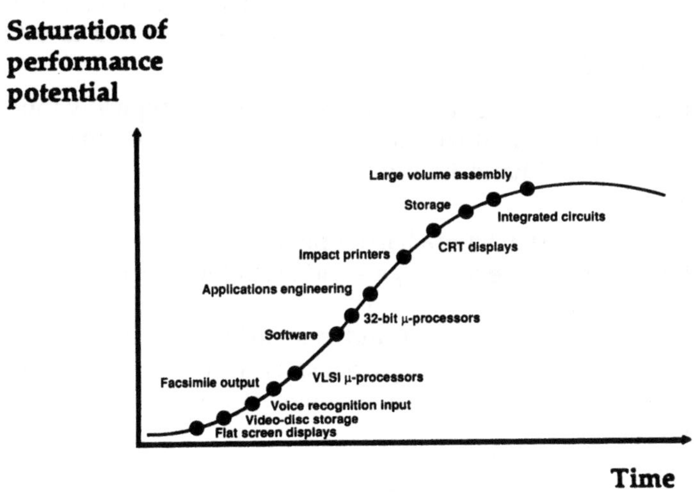

Figure 7: Companies have to watch a wave of technologies

their equivalents in the USA can achieve through the use of such partnerships.

Technological leadership has to be shared, and the mechanisms for shared technological leadership have to be cultivated and managed. This has been one of the key findings of a worldwide Arthur D. Little investigation of how successful innovative companies tick.

Lessons for Small Companies

The typical entrepreneur of a small technology-based company underestimates the cost of penetrating the market and the acceptance barriers against newcomers. He believes that the attractive technical solution provided by his product speaks for itself.

However, technical and market barriers are growing rather than diminishing.

- Switching to new technological solutions increasingly means that customers have to invest in changes in their whole system. In order to introduce computer-aided design (CAD), for example, a company has to link it to its computer-controlled machine tools as well as to its commercial data processing system.
- There is a dual risk in introducing new solutions from a small firm. Will the new solution work and will the supplier have the resources and staying power to support his products over time and to offer back-up in case of technical problems?
- The decision processes in established companies are typically much longer and much less organized than those estimated by small suppliers. Marketing and hand-holding take a lot of presence and perseverance.

As a result, small firms often have a hard time convincing their potential customers of their innovative offerings – and often they do not have the financial muscle to wait.

If we look at many of our high-tech and innovation-driven companies in Europe, many of them in technology parks, some of them supported by venture capital firms, their main problem is not technology, but marketing and sales.

We therefore believe that theregional national and European technology sponsorship programmes which have blossomed in recent years are missing the point. What Europe needs now is to activate the third innovation channel, i.e. help in getting technologies to the market.

The Cooperation Phenomenon leading to "one plus one equals three" which works so well in the USA and in Japan, has to be brought to the attention of entrepreneurs and executives of our large European corporations.

How can this be done?

How to Marry Technological Dynamism and Marketing Capabilities?

We need mediators in order to trigger more strategic alliances between small technology-based firms and large established corporations. The objective is to complement the different strengths of these two types of players in the innovation game to achieve economies of scale for investments in innovation, to overcome barriers against technological advances and to strengthen our European economies in the international competitive arena. This is a particularly challenging task if we want to activate the third innovation channel across traditional boundaries in Europe.

Who can be the mediators?

In a first stage, we believe that governments and the European Community need to focus on encouraging strategic alliances particularly between young technology-based firms and large established corporations.

In a second stage, we believe that it is in the interest of the European financial industry to mediate between their customers from both camps. In fact, there are first attempts of this kind in Berlin, where 5 leading banks (Deutsche Bank, Dresdner Bank, Commerzbank, Berliner Bank and Stadtsparkasse) have come together to set up a Foundation the objective of which is just this: to help establish cooperation between small technology-based firms and large corporations. Some of the large German corporations, e.g. Siemens and Schering have joined this Foundation. Arthur D. Little is also a member of it, because we see a benefit for management and technology consulting firms to be a mediator. In many cases it puts us in a position to help our large industrial clients get access to new technologies.

In a third stage, large established companies should get organized and have their own mediators (as we can observe them in the USA and in Japan) – a "Mister Entry" whose task it is to be the first contact point for small technology-based firms and to ensure that they are being coached and dealt with carefully within the organization. Their main concern is to overcome the NIH-syndrom (NIH = not invented here) and the behavioural resistance to innovative thinking. At the same time, their advice can make the small technology-based firms much more effective in penetrating the large organization and in gaining the respect they deserve.

To be successful the third channel for innovation needs such a "Mister Entry" to mediate the partnership process.

Constructive Cooperation between Smaller Firms for Efficiency, Quality and Product Changes

P. Mariti

Introduction

Dating roughly from the beginning of the 70s, there has been increasing awareness among scholars, firms, research agencies and governments of the fact that the structure and characteristics of inter-firm complementary, auxiliary and subsidiary relationships – i.e. the quality of "constructive" interdependence of business organizations – really matters for regional and national development and for international competitiveness as well.

Initially identified in relation to large and very large businesses, particularly in "high technology" sectors, and in the world of large firm to small firm links[1], such relations have subsequently been recognized in the world of smaller firms as well. What follows deals mostly with small firm to small firm cooperation. Much of the same analysis can also be extended to cover the large to small firm area, with the proviso that one of the crucial problems here relates to whether and how small specialized firms can protect themselves from large buyers' (not necessarily large firms') attempts to exploit the relationship unilaterally. As to large firm to large firm relations, these may have peculiarities and certainly have overall impacts of their own.

Here the focus is mainly on a general positive analysis of the economic and technological forces underlying the phenomenon that Alfred Marshall himself christened "constructive cooperation" as opposed to individualistic behaviour by firms and/or behaviour aiming "at maintaining prices at higher levels, than would otherwise prevail"[2]. Organizational and strategic aspects as well as the issues raised by the evolution from simpler to more complex forms are not addressed. They are very important for networks in which one firm has a "focal" or "central" position (as is typically the case with a franchising network) and for groups of firms which are basically under the same managerial control ("internal linkages")[12].

1 Surveys can be found in [5], [29].
2 See [20], book III especially chapters XI and XII.

The Firm as a Set of Activities

A suitable vision of the firm should first be developed. An input-output table provides a rough picture of the intricacies of a contemporary industrial system from the point of view of firms as buyers of inputs and sellers of outputs. The implication is that the production of any given commodity is in fact brought about with contributions from many firms. Assuming this rather obvious fact as the starting point for analysis, it will then be convenient for our purposes to go further and look at firms as entities undertaking and organizing a certain number of different activities rather than buying inputs and making marketable products according to a given production function. In other words, the firm is partitioned not among the markets in which it buys its inputs and sells its outputs, but rather among the activities that constitute the scope of its overall undertaking: the firms is thus viewed as a set of activities (or processes) which by their very nature take place over time.

As used here an activity is a primitive concept and cannot be defined too closely. Some of its characteristics need be pointed out however:

a) activities may be as diverse as those relating to search and assessment of consumer wants; research and development; product design; execution and co-ordination of processes of physical transformation; marketing of goods; searching for financial resources, for partners or for a suitable location, and so on;

b) each activity leads to an "output" which is potentially marketable, though it may be difficult to define the "output" of activities like research and development in physical terms;

c) activities take place within the firm, though not to the same extent in every firm[3];

d) activities are functionally related in the sense that each adds value to the final or marketable product(s).

It is of considerable interest to note that in the field of business economics and accountancy scholars are currently trying to find new solutions capable of representing the true contribution of each single activity of the firm to the product cost. The reason is that traditional accounting has many drawbacks particularly in times of rapid innovation. They speak of "activity costing" and focus on an analysis of the firm

3 The internal organization of smaller firms is often understandable rudimentary as to the specification of activities.

which aims at distinguishing between "resources, activities and products"[10].

A Plausible Analytical Framework[4]

Firms exist to provide products and services and their main aim is to stay in the market and to prosper or at least survive. To achieve this aim they vie with their direct competitors by: 1) striving to secure lower total costs than rivals in facing price competition; 2) making their products attractive via quality and product changes (nonprice competition); 3) attempting to get more resources with which to work. What follows deals basically with these three lines of action by firms.

Static Efficiency

Consider firm A as undertaking various activities with a cost function associated with each of them and aiming at minimizing costs. For the sake of simplicity, let us concentrate on a single activity among the many carried out by the firm. It could be an intermediate stage of production or, in the field of immaterial production, the activity of searching and opening up new foreign markets. If the activity is characterized by economies of scale, there will be a tendency for a firm to dispense with the activity under examination (if relevant costs are negligible) and buy the required level of "output" from an outside specialized supplier, firm B, thus fully exploiting economies of scale and producing at unit costs lower than those obtainable internally. It is worth remembering that the existence of economies of scale does not necessarily depend on large plant size: they can often be fully exploited at levels of output that represent only a small share of the total market and thus small firms can be efficient. Indeed, many such examples can be found [2]. Besides, the significance of unit cost reductions is measured by the gradient of the scale curve, which may allow efficient operating even at scales smaller than the minimum efficient one.

If the relevant intermediate market is competitive, the price paid by A to B will be lower than its internal costs and total final product costs will be reduced accordingly. Since demand for the "output" of the activity under consideration is derived from that for the final product, if the indicated tendency is to take place the demand for the final product

4 The next two major headings are a summary and, within present limits, an updating of [17]

must be growing in real terms – so as to make the activity in question profitable on a separate and stable basis for firm B.

This is the well known process of division of labour between firms first analyzed by Adam Smith and recently by others [33] [28] [1]. That this tendency is at work is evidenced by the clusters of firms in the many so-called industrial districts in Italy as regards such sectors as furniture, leather and shoes, textiles, light mechanical engineering and component makers supplying the car industry in most countries. The emergence in the more industrialized nations of a wide range of firms specializing not only in particular technologies such as lighting, cooling, ventilation but also in such broad fields as business organization, advertising, data processing, market research and so on can be explained along much the same lines [22].

Economies of Grouping Different Activities within The Firm

Some of the further potentialities of looking at the firm as an organized set of activities can be assessed by referring to the more conventional treatment of the management problem in economics [31]. An increase in the activities organized within the firm tends to produce strains within the administrative structure which raise the costs of organizing additional operations (even identical to those already undertaken): the rise in cost occurs both because the administrative costs themselves rise, and because there is a control loss. Those making decisions make more mistakes and fail to allocate resources wisely. The presence of a given activity may significantly lower the cost of others. A given set of activities can facilitate the organizing and performance of some processes but hinder those of others.

Indeed, most entrepreneurs would stress that the cost of carrying out an activity within any given firm is seldom independent of: 1) what other activities it is engaged in, 2) their number, and 3) their nature[5]. As regards adding a new one, a top executive has been quoted as saying: "For purposes of making the decision to add an activity which may be fairly permanent, the schedule of estimated costs of production should include not only an exhaustive list of the probable added costs but also a complete assignment of all of the overhead charges (both fixed and variable)."...since..."the added activities will encroach upon available

5 Preliminary findings of ongoing extensive research by the present author are much line with these propositions.

capacity within the firm and this will lead to an actual, though unanticipated and perhaps unrecognized, increase in costs"[2].

These are some of the reasons why firms look for selective rather than uniform expansion and development of their set of internal activities. But there is more to it. Successful activities strengthen firms and unsuccessful ones weaken or ruin them. The full potentialities of conceiving of the firm as a set of activities can be gauged by pushing the concept further along the lines originally suggested by Richardson [24]. Activities within a firm are basically of two types, quite independent of economies of scale:

a) *similar* activities, in the sense that they require the same capabilities (knowledge, experience and skills) for their undertaking. A capability may depend upon command of some particular technology in the field of mechanical engineering, electronics or food processing, or may derive from skills in marketing or knowledge of and reputation in a given market. Firms enjoy economic (and non-economic) comparative advantages and cumulative learning relative to their competitors;

b) *complementary* activities, which are separate phases of the overall production process not requiring the same capabilities, but nonetheless, having in one way or another to be coordinated to reach the stage of final production. It is important that this notion be understood to describe not only the relationship between the manufacture of final products and their components, but also the relationship of each of these to the corresponding activities of research and development and of marketing.

Examples may help to clarify the definitions: brick and cement-making are complementary to house building, but they are not similar to it; the ability to market commodities may be similar in such fields as ballpoint pens, safety razors, shoes, carpets, etc. but the activities to turn out these goods are not complementary; the production of clutch linings is complementary to that of clutches and cars, but in that they are made by firms with a capability in asbestos fabrication, they are similar to drain-pipes and heat-proof suits; the capability of catering may be similar for weddings and other social gatherings (and in some cases even for non-food catering), but it is not similar to that of basic food production and processing.

If activities are separable firms will normally be able to add or expand some of them and to contract or discontinue others. Complementary activities need not and in general will not be similar. In their search for selective rather than uniform expansion, firms will tend to specialize in a similar group of activities and to rely on purchasing from outside for others to an increasing extent. The tendency will be stronger when activities that are complementary within the original set show significant increasing returns.

Firms will be both more efficient and more effective (i.e. able to sustain their market competitiveness) the more they can concentrate on similar activities, while buying outside the required levels of "output" of activities that are complementary to their final product(s), but not similar to those organized internally. To put it differently, since activities which are complementary for one firm may be similar for another, they are led to complement each other in terms of how their resources are specialized by activity. Exploiting this complementarity often requires very close linkages.

Economies of scale are a *centrifugal* force for a firm that may lead to "productive decentralization" and to so-called "flexible specialization". The capabilities of firms and the related economies of grouping similar activites have instead a *centripetal* effect: firms tend to organize their undertakings around their capabilities to fully exploit comparative advantages and dynamic economies (or "learning by doing"). A firm's capabilities, and thus its relations with other firms, expand and alter over time in response to economic stimuli, and to technological and environmental change[6].

The often impressive and ever-changing mosaic of firms that can be observed at the industrywide level, or in a given industrial sector or industrial region can now be better understood: a particular (or different, more efficient and effective) grouping of activities within a given firm involves a particular (or different, more efficient and more effective) distribution of activities between firms (and vice versa). An important result in this context is that a firm's propensity and ability to forge external linkages depends to a significant degree on its internal activities and resources [25]. No firm is an island: they do tend to form archipela-

6 The concept of economies arising from "grouping" similar activities within the same firm, i.e. economies obtainable by concentrating on activities where the firm has accumulated knowledge, experience, skills and other resources, is not the same thing as the comcept of ecnomies of "scope": the latter refers to products and is thus typical of multi-product firms; the former applies to the one-product firm as well. Smaller firms are well advised to avoid fields where there are dominant economies of scope which large firms may be willing to exploit.

gos keeping in touch with other firms to secure the outputs of their respective activities and engaging even in complex external linkages.

Clusters of relationships do not necessarily imply much cooperation between the firms involved: many items and component parts can be bought from catalogues, or via spot transactions, subcontracting, long-term contracts, etc. Several services can be leased. Many relationships between firms are well mediated by markets and prices. Indeed, the essential feature of the base economic theory is not to deny that there are interactions between firms. Rather, the contacts or linkages considered are those organized *exclusively* by the use of the price mechanism; firms behave in isolation once the appropriate price signals are given (and their behaviour may simply be added together)[11].

The interest here is in relations between firms that are not mediated or organized, at least directly and/or in the short and medium run, by markets and prices.

Informed Markets and Product Quality

Life would be easier for firms were it not for the inability of some markets to provide goods in sufficient quantities, if at all, and with the right quality characteristics. The causes underlying market failures have much to do with the contending (and often conflicting) wills and property interests of the firms, information and transaction costs. The quality of the goods (and services) to be exchanged and the validity of the rights to use them in specified ways must be checked. There are costs in arranging transactions: offers must be communicated and alternatives compared. Checking prices, closing contracts, verifying their execution, collecting payments, guarding against fraud or other non-performance, assessing competence and reliability of suppliers are all costly activities. Information is not something free and is in general asymmetrically distributed.

To get at least the flavour of the factors at play and which are conducive to the emergence of constructive cooperation between firms as a rational and purposeful action, it is convenient to go in some depth into a specific problem – that of product quality. The choice is not casual since the quality issue is at the heart of all product innovations. Cooperation is also a matter of degree. There is almost a continuum between "soft" cooperation and full coherent constructive action implying a common project around shared interests over a protracted (though finite) time horizon. Several examples of cooperation and networking can be brought to light.

Markets for industrial products and services can generally be considered as informed markets, that is markets in which the average buyer knows enough about the quality of goods and services offered to appraise them on their own merits. Firms being professional buyers can more easily detect poor quality or fraud and thus need less to rely on brand reputation, advertisements, warranties, etc. to obtain information about quality. It is worth noting that we can expect that, all other things being equal, small sellers will face fewer disadvantages when buyers are expert.

Nevertheless, specifications, details, instructions and the like are often very important in intermediate markets. They generally make sense for large purchasers since:

i) there are high fixed costs of specification (learning what aspects are crucial, writing the specification, testing the product etc.);

ii) they invite competitive bidding;

iii) less reliance needs to be placed on the integrity of an individual seller and the risk is smaller that the quality will change with subsequent purchases.

The precise details of the specification help eliminate much of the fuzziness in exchange: a buyer receiving an inferior product can frequently claim breach of contract rather than merely registering a complaint. All this helps to explain why, whenever possible, groups of large firms are ready to negotiate and write industrywide standard specifications: they reduce the user's search and coordination costs.

Sellers also gain advantages from collective action. Professional buyers are sometimes able to discern average product quality within an industry, but not the quality of a particular item. An individual seller may find it too costly to distinguish his product: the sellers may be too small and potential customers too scattered for a major attempt at product differentiation to be profitable. In such circumstances, an individual seller has little incentive to upgrade his product since benefits accrue to the industry as a whole. Indeed, the individual incentive of any seller may be to decrease quality and cut costs.

As a group, however, sellers sometimes have a strong desire to raise average quality and thus a powerful motive to act collectively to promote quality improvements, often under the auspices ofgovernmental or local authorities. Firms exporting into expanding markets for high-quality items provide examples: by acting collectively, small or medium-sized firms are able to achieve economies in guaranteeing product quality that is already possessed by the large-scale enterprise. One way of exploiting further network externalities is to complement the collective action for quality with recourse to other national suppliers offering complements which are both inexpensive and of high quality, as this increases overall demand: food products, beverages and pottery;clothes, leather products and accessories provide examples of items manufactured and sold by varying industries but used conjointly in "package" offers by consortia of firms comparable in size and expectations. Incidentally, this may also generate returns from the point of view of national exports, justifying supporting actions by public authorities, since "averaging" externalities may also occur between industries in the sense that, for instance, a good product from Italy may make importing countries more willing to buy other Italian products as well.

National cooperation between firms can be complemented with that at the international level: knowledge possessed by local firms about tastes, standards of sameness (allowing for interchangeability of the products of different manufacturers), safety standards, identity standards, etc. can ensure and/or speed up introductions to the relevant markets; if use of a local distributive network is made possible as part of an agreement, enormous direct investment savings are also possible. Entry barriers for smaller firms are significantly reduced in this way.

Overcoming Absolute Market and Internal Limitations on Product Change

Even casual observation reveals that firms under innovative pressures in informed industrial markets typically reflect a variety of tastes and a taste for variety as a consequence of market requirements for product change. This creates demand for a whole range of different types and models of the same commodity; but, at the same time, it limits the demand for any particular type or model within the range. At least two problems may arise:

a) very few sellers are willing to accept the offer; this may lead to a "small numbers relationship" that might cause lengthy and costly bargaining over the terms of trade, and in some cases the parties may not be able to reach agreement at all;

b) the supplier firm may have to designate "fixed" resources to meet the requirements. The meaning of "fixed" here depends on the specialization of resources in the firm: the manpower and machinery employed, buildings occupied but also product designs and the research and development to enable it to produce more cheaply and/or meet other firms' requirements may be so specialized as to be of little or no use in alternative allocations independent of transaction costs. In other words, they may have little or no resale value and thus the firm may face losses or "sunk" costs in the disposition of such specialized resources.

The presence of such "fixed" or immobile assets brings a degree of irreversibility into the firm's allocation decision. Firms must be very cautious where specialization of their resources is involved and be sure that there will be a stable demand for the output of their activities. To the extent that demand can be regarded as being permanently stable or even growing, on the basis for instance of an agreement with the client firm, it will become increasingly attractive for a firm to allocate additional "fixed" specialized resources. But then transaction difficulties (and costs) may arise: whenever there are specialized and durable resources which cannot be enlarged or reduced overnight the remuneration of factors contains a surplus or quasi-rent in excess of its cost. The extent of this is precisely determined by the level of demand for commodities affected by such difficulties and firms may be tempted to appropriate these additional earnings, i.e. to behave unfairly ("opportunistic behaviour"). The problem is exacerbated if there is learning involved in the relevant activity: by its very nature it can be acquired only over time and a buyer firm may find it difficult to add the relevant activity or even resume an activity dispensed with previously. On the other hand, switching to another supplier may not be feasible in the short term and anyway is costly.

There are cases, indeed, in which careful formal specification of desired designs, characteristics or adjustments is very costly and time consuming or even impossible to achieve altogether. This is most likely to happen when:

a) intangible quality aspects such as "workmanship" are at issue;
b) technological change in the relevant field(s) is fast and/or not easily foreseeable and/or of the across industries type (as is the case from electronics to mechanical engineering);
c) there is need for frequent product changes and rapid innovation for a firm to stay in known but turbulent markets;
d) rapid product adjustments are required to serve new expanding but unknown markets, particularly distant foreign markets.

Under such circumstances firms may be desperately seeking information on the quality of the items or services they need before they can confidently use them, and also on the technical competence, delivery and service reliability of suppliers, on local "goodwill" and the financial strength of clients in distant markets, etc. Furthermore, ordering by unusual specifications or ill-defined specifications – because, for instance, the required items are produced according to a technology which is alien to the one over which the firm has command – takes time and there are risks of very costly misunderstandings, disputes and further delays.

The realistic and feasible alternative is not the "incremental" one suggested by the markets and hierarchies approach [32], making a "substitution at the margin"[7] of the market with internal production when the cost of using a product is higher than the cost of organizing the required transactions[7]. On the one hand, recourse to the market is often not possible for implementing some innovations because of its inability to provide the most needed specifications and designs and the necessary adjustments at the right times[13]. It is not a failure of the market to achieve perfection, i.e., a relative failure, but an absolute failure with exchange costs being infinitely high. On the other hand, activities necessary to secure all that is required may be dissimilar or outside the scope of the firm through, for instance, mergers, acquisitions or "internal linkages".

The real outcome in such cases would very often be the curtailment or postponement or even the abandonment of the whole innovative idea. Cooperation thus emerges as a viable organizational arrangement that firms can use to retain all the advantages of concentrating on similar activities (reaping related economies)and keeping transaction costs as low as possible, while getting precisely what they need both quantitatively and in terms of designs, specifications and quick adjustments. Looked at in more strategic terms, conscious cooperation implies that what are usually regarded as exogenously given structural characteristics of market transactions (small numbers, bounded rationality, assets specificity, etc.), moulding firms interactions often in ways det-

7 Activities are organized groups of single transactions.

rimental to their true needs and competitive requirements, can be kept under control endogenously by firms themselves [26][8].

When different specialized contributions are clearly perceived by firms as complementary to achieve product quality and product changes *allowing* a stable or growing *demand for all*, short-run profit maximization types of behaviour and opportunism are superseded by constructive behaviour. As Marshall put it "...mere cooperation can attain...many of the ends for which fusion is sometimes sought; and without lessening the scope for initiative by independent firms." The last part of this statement reminds us that, in addition, the market test is still applicable. Nothing can prevent either party from making alternative arrangements. This retained discretion not only helps ensure that mutual benefits continue, but also results in lower costs for the buying firm [14].

Some Facts about Small Firm to Small Firm Cooperation

The foregoing general analysis can be partially complemented with some facts. On the evidence to date cooperative linkages, in terms of their nature and frequency, are only rarely or loosely related, if related at all, to the industrial sector or thetechnological base of the firms concerned. This permits only qualitative statements mostly confined to broad aspects.

The treatment is far from exhaustive and reflects the state of infancy of the subject (and the Italian base of the present Author as well). It is perhaps a fair assessment to say that nonetheless it confirms some "abstract" reasoning and also much anecdotal evidence.

General Aspects

As to the influence of size, while there are no doubts about the diffusion of cooperative agreements at the national level, research conducted on a "sample" of firms in sectors reflecting the characteristics of the Italian industry has reached the conclusion that small size does not represent an absolute obstacle even at the international level [4]. Another piece of research relating to small Italian firms belonging to " high-technology" sectors reaches basically the same conclusion with an interesting qualification: cooperation with European firms is much higher relatively than

8 Japanese authors stress the role of trust as a basic ingredient of cooperation. Trust is largely a product of culture and depends on the value preferences of a given society, In this paper the term "cooperation" means the rational and purposeful response fo firms to economic and technological stimuli in all cases in which there is no other viable alternative.

with overseas firms; greater knowledge of geographically close markets and greater cultural homogeneity seem to facilitate firm linkages [27][21].

Cooperation between small firms tends to be confined to single activities and/or single products and/or single markets in the sense that global alliances are rare. However, the content of the agreement tends to be "complex" (more than one flow of products, components or technology between the company up to the creation of common structures), though without reaching a great financial or managerial involvement as compared to what happens in the large business sector. As we shall see presently, these are important aspects for agreement success.

Equity agreements (joint ventures or exchange of shares) are found in a minority of cases; contractual agreements (formal and informal) represent the largest quota. As to broad areas of firms' activities, the commerce or market area largely prevails (motives: market quota expansion or consolidation of competitive position, entering new markets, complementing product range, etc.). Agreements for production rationalization and to reap economies of scale are also important. The field of technology (sale, acquisition and exchange) appears less important and seems confined to technology applications more than generation or development. R&D cooperation activities score the lowest positions. It is important to point out that these results may reflect the peculiarities of Italian industry, though a deeper and more general explanation may be that R&D – i.e., innovation plus learning – is in fact the outcome more than the content of a specific agreement. Firms may look for linkages even in distant areas such as marketing to get feedback knowledge leading to much product change later on. Innovative companies and companies operating in "advanced"sectors are often found to entertain more external contacts than others [25][11.1]. This is surely a sign that such firms have higher requirements than those operating in sectors demanding lower levels of innovations. With reference to small UK biotechnology firms, R&D activities seem to be carried out through a certain range of linkages with other firms more than by specific formal R&D agreements requiring firms to work closely together on sizeable projects [30].

As to sectors, very recent research confirms previous findings [9]: cooperation between smaller firms is found mostly in relation to "traditional" sectors (such as textiles, garments, leatherwear, wood and furniture, etc.) and in sectors requiring the assembly of many items (machine-tools, agricultural machinery,light engineering in general, car components, etc) and much less in process industries (paper, board, food, beverages,etc.). Stating that cooperation between smaller firms has been observed mostly in relation to "traditional" sectors does not imply that it is not common or of less importance in other sectors as well.

More Specific Advantages

Beyond the general advantages that have been discussed, there is wide agreement among researchers regarding a certain number of more specific advantages, whose importance will vary with the sector and the firm activity involved. As to production and investment, cooperative relationships can:

a) ensure a greater flexibility in and for firms in the specific sense that less fixed commitments are necessary in assets and activities which are alien or not essential to the core of their undertakings. This in turn implies lower investment costs than direct investment with lower exit barriers. The outcome is a greater ability: 1) to adapt to both foreseen and unforeseen changes in the level and composition of demand and 2) to re-design products and manufacturing processes in the face of new competition;

b) facilitate the necessary closeness between users and producers. The ensuing two-way relationship is essential to all product innovations, both as processes and as single events [13].

In marketing terms, being close to firms serving final customers provides knowledge of even distant market conditions thanks to the market familiarity of local partners. This may be crucial: 1) in accessing distribution channels before others, since cooperative linkages can allow a greater speed of introduction to markets in the first instance relative to direct investment; 2) in finding or adjusting to new markets, since local firms may be able to provide information (on trends, quality, compatibility, safety standards,etc) about such markets not otherwise obtainable by small firms located in distant areas; 3) in expanding the distribution network with low cost investments; 4) in reducing the costs of product differentiation or of launching new products.

In the field of R&D and the exchange of technology and information the most common advantages are: joining complementary strengths; risk sharing; obtaining relevant information; the possibility and speed of introduction to an unknown technology (time saving); obtaining additional revenue through the sale of technology. One cannot but entirely agree with the following statement by Rothwell and Beesley summarizing the issue: "Innovation consists of much more than just solving the technical and physical problems associated with product design and development. It consists also of a whole range of activities prior to the actual act of physically producing the new product. For a small firm, developing a new product can be a very risky activity. It demands the commitment of a large proportion of investment funds to a single project, together with the opportunity costs involved in committing scarce skills and time to one activity rather than another"[25].

On the Scope of Cooperation for Smaller Firms

Cooperation is a fashionable topic and of course one should beware of ideas which may be just a passing fad. Cooperation is not a panacea. There are some drawbacks to it that may lead to unsuccessful experiences and limit its scope for smaller firms.

The risk of losing independence should not be underestimated when the relationship is between a large user and a smaller seller. A large buyer is not necessarily a large firm by any dimensional variable (capital employed, sales, employees, etc), but rather should be taken to mean the firm that absorbs a significant quota of a supplier firm's output, so that if the buyer removed its business from that particular supplier, he would be in serious trouble. Lack of capital and funds for current expenditure makes smaller firms vulnerable to domination by established enterprises [30]. Larger firms may promote cooperative agreements with smaller ones in order to keep possible sources of technical progress under control and to discourage entrants to the market. On the other hand, smaller newcomer firms may enjoy considerable advantages from a cooperative relationship with large established firms and gain credibility in the wider market as reliable suppliers. It is worth remembering that situations where small firms have problems with, behavioural alternatives to and forms of resistance regarding too tightly specified or too restrictive requirements on the part of large firms have long since been clearly analyzed in realistic terms [3].

Leaving aside the question of profit sharing which is probably less relevant than some economic models would lead one to believe, the sharing of proprietary technology and R&D results may entail the risk that partners will use it covertly or in ways not agreed upon. The risk of spillovers to other firms is also increased. The design of contracts is critical but small firms are ill-equipped for writing contracts. In addition, it is usually very difficult even for large businesses to get full protection or to foresee the future development of the relationship. This last point reminds us of the fact that a relationship is never static but develops over time: a partner may reveal himself to be as suitable as was initially expected or may fail altogether. There are no easy ways out of these problems with the aggravating factor that formal contracts may turn out to be a barrier to exit.

A related point is that of difficulties in properly managing the progress of the agreement even when partners live up to it. This may result from the fact that the circumstances that initially motivated it have changed somewhat in the interim. Difficulties are more likely to arise when firms share too many activities. On the prescriptive level, small firms would be well advised not to form very demanding agreements which require deep involvement, especially in terms of financial and managerial resources.

There is also the danger of clashing in the same market once the agreement is over. This may provide part of the explanation why horizontal agreements are rare. Such agreements are more likely to be concluded when according to the agreement both firms will sell their partners products and will introduce a series of technical adaptations and modifications which will make them suitable for the other's market.

Though much more research is needed before sound statements can be made, it is interesting to note that several authors see cooperation as especially important for innovative firms. In addition, some research findings point to the fact that cooperation is very effective in vertical structures; it is not by chance that it has actually been tracked and deeply studied in the context of user-producer relationships, particularly inter-area subcontracting by Japanese researchers, and of large firms to small firm relations, particularly in car components. Some evidence suggests that it is much less likely horizontally: in the field of industrial automation, production firms located in Lombardy (Italy) did not reach the stage of "supply systems" in spite of the presence of favourable side conditions such as the "systemic" nature of technology, complementary firms' technology, etc.[15]. A very preliminary conclusion could be that cooperation is most suitable for gradual, step-by-step innovation and the quick adjustments required by changing markets rather than for major innovations and/or process innovation. R&D cooperation activities should thus be confined to technology that is advanced only by small scale research projects directed at the particular problems of that industry. This should not be taken to suggest that gradual innovation, even perhaps of the most trivial or purely imitative kind, is not important or is of lesser importance: minor innovations are crucial since they are cumulative over time and tend to eventually produce major product changes and innovations.

Cooperative linkages – as organizing and coordinating mechanisms – give considerable market and organizational flexibility to firms and groups of firms. Current developments in manufacturing flexibility through Computer Integrated Manufacturing (CIM) seem to provide a technological solution to similar problems. These techniques generally imply substantial investments, often of a rather specific type. By using them, larger businesses can change products on the assembly line quickly and thus serve different market segments or niches at a time and at lower costs than smaller firms. Flexible manufacturing techniques could act to negate some of the advantages of smaller firm cooperation and reduce its scope. There is some preliminary evidence indicating that vertical relationships between firms are reduced by some advanced forms of flexible automation applied to design and production activities [16]. "Centripetal" tendencies depending on integrated automation should not be underestimated and smaller firms should avoid areas where flexible manufacturing techniques are developing. With reference to Japan, there is however some support for the idea that CIM may in fact be restricted to single applications, which nevertheless are important in solving specific production problems [23].

Under concomitant pressures of market expansion (or enlargement as a consequence of a reduction in barriers to trade) and technological change, smaller production firms will be pushed towards greater specialization. Their external needs, both technical and otherwise, will increase. They will require many accompanying service inputs to be effective in the market place, such as information services, managerial consultancy, quality control, contract engineering, product servicing, manufacturing management, etc. There is much scope for cooperation between smaller "manufacturing" firms and "servicing" ones in all cases where customizing designs and products are important and in knowledge-intensive or science-based sectors (biotechnology, information technologies,new materials, energy). Incidentally, service industries are by their nature less likely to experience so-called "global" economies and thus it can be expected that "global" firms will stay out of such sectors.

The establishment of worldwide networks of sales and service subsidiaries in some sectors may be crucial for smaller operating firms in those areas. There is much scope for interfirm cooperation, organized at the national level and aimed at the international level, given the high cost involved in relation to firm size in accessing international markets. [8].

To conclude, cooperation is not a substitute for leadership and foresight in the management of small firms and to be successful often requires the contribution and converging efforts of other institutions such as universities, banks and public agencies at the local, national and international levels.

References

[1] Ames, E., Rosenberg "The Progressive Division and Specialization of Industries", *Purdue Faculty Papers in Economic History 1956-1966*,**Richard D.Irwin, 1967, pp.345-362**
[2] Blair, J.M., *Economic Concentration - Structure, Behavior & Public Policy*, Harcourt Brace Jovanovich, New York, 1972
[3] Blois, K.J. "Vertical Quasi-Integration",*Journal of Industrial Economics*, vol.20, n.3, pp.253-271, July 1972
[4] Cespri, *International Agreements of Italian Firms: an Analysis with Specific Reference to the Role of Intra-EEC Cooperation*, Universita' L. Bocconi, Milan, March 1988
[5] Chesnais, F., "Technical Cooperation Agreements Between Firms", *STI Review*, vol.3,n.4., December 1988, pp.51-119
[6] Coase R.,"Industrial Organization: A Proposal for Research", in V. Fuchs (ed.), *Policy issues and Research Opportunities in Industrial Organization*, NBER, Columbia Univ.Press, New York, 1972, pp.59-73
[7] Coase, R., "The Nature of the Firm", *Economica*, n.4, November 1937, pp.386-405
[8] Dalum B., "Interfirm Cooperation Agreements - Two Cases from the Danish Electronics Industry", Paper presented at the EC/FAST Seminar on Interfirm Cooperation Agreements, Paris X-Nanterre University, 21-22 November 1988 (mimeo)
[9] Jarrillo, J.C., Martinez, J.I., "Evidencia empirica de la superioridad de los sistemas cooperativos", *Economia Industrial*, n.266(Marzo-Abril 1989),pp.99-104
[10] Kaplan, R., Cooper, R., "Measure Cost Right: Take the Right Decision", *Harvard Business Review*, vol.27, n.3, Sept.-Oct.1988, pp.96-103
[11] Kirman, A., "The Intrinsic Limits of Modern Economic Theory: The Emperor Has No Clothes", *Economic Journal*, vol.99, (conference), pp.126-139
[11.1]Lassini, A., *Opportunita' tecnologiche, piccola dimensione e strategie innovative*, Il Mulino, Bologna, 1986
[12] Lorenzoni, G., Ornati, O.A., "Constellations of Firms and New Ventures", *Journal of Business Venturing*, 3, 1988, pp.41-57
[13] Lundvall, B.A., Product Innovation and User-Producer Interaction, *Industrial Development Research Series*, n.31, Aalborg University Press, Denmark, 1985
[14] MacMillan, K., Farmer, D., "Redefining the Boundaries of the Firm", *Journal of Industrial Economics*, vol.27, n.3, pp.277-285, March 1979

[15] Mariotti, S., Riva, A.(ed.), *Il settore dell'automazione industriale in Lombardia.Strategie di competizione e collaborazione tra imprese produttrici,* Angeli, Milano, 1989
[16] Mariotti, S., "Il riordino della funzione tecnica di produzione dell'Industria", *Politecnico di Milano,* Rapporto n.89.030, Aprile 1989
[17] Mariti, P., *Sui rapporti tra imprese in una economia industriale moderna, Angeli,* Milano, 1980
[18] Mariti, P., Smiley, R., "Cooperative Agreements and the Organization of Industry", *Journal of Industrial Economics,* vol.21,n.4, June 1983, pp.437-51
[19] Mariti, P., "Los acuerdos de colaboracion entre empresas en las economias modernas: Algunos apuntes para la reflexion", *Economia Industrial,* n.266 Marzo-Abril 1989, pp.61-69
[20] Marshall, A., *Industry and Trade,* MacMillan, London, 1919 (second edition)
[21] Montebugnoli, E., Schiattarella, R., "Gli accordi internazionali tra imprese: considerazioni su soggetti e motivazioni con particolare riferimento alle piccole imprese", Ottobre 1989, (forthcoming)
[22] Petit, P., La Croissance Tertiare, *Economica,* Paris, 1988
[23] Piacentini, P. "Note sullo stato e le prospettive di sviluppo del l'automazione flessibile nell'industria manufatturiera giapponese", 1989 (forthcoming)
[24] Richardson, G., "The Organization of Industry", *Economic Journal,* vol.21, n.1, pp.883-896
[25] Rothwell R., Beesley M., "Patterns of External Linkages of Innovative Small and Medium-Sized Firms in the UK", *Piccola Impresa/Small Business,* vol.2, n.2, 1988, pp.15
[26] Sako, M., "Comparisons of Sub-contracting Relations in Britain and Japan: an Illustration of How Trust Affects the Way Firms Deal with Risk and Uncertainty", Third Workshop on "A long run perspective on structural change and industrial organization: a comparative study of Italy and Japan", CNR, Siena, May 1989, (mimeo)
[27] Start, *Indagine sulla cooperazione produttiva e tecnologica tra imprese europee,* Roma, 1988, (forthcoming)
[28] Stigler, G.J., "The Division of Labour is Limited by the Extent of the Market", *Journal of Political Economy,* vol.59, n.3, 1951, pp.432-49
[29] Viesti, G., "Gli accordi di collaborazione fra imprese: nuove strategie per la crescita internazionale e l'apprendimento tecnologico?", *Cespri,* n.11, Milano, 1988.
[30] Walsh, V.,"Desperately Seeking Solvency: or External Linkages of Small UK Biotechnology Firms", Workshop on Inter-firm Technological

Co-operation Agreements, Universite' Paris X-Nanterre, November 1988 (mimeo)
[31] Williamson, O., *Corporate Control and Business Behavior*, Englewood Cliffs, Prentice-Hall, N.J., 1970
[32] Williamson, O., *Markets and Hierarchies - Analysis and Antitrust Implications*, The Free Press, New York, 1975
[33] Young Allyn, "Increasing Returns and Economic Progress", *Economic Journal*, vol.38, n.152 December 1928, pp.527-541.

The Small Firm Perspective on Inter-firm Collaboration for Innovation

K. Dickson, H. Lawton Smith and S. Smith

Introduction

The Small Firm in the Innovation Process

Technically specialised small firms are becoming increasingly important in the innovation process, particularly as collaborators with larger firms where their specialised knowledge may be critical in the development of new "hi-tech" products. Small firms have long been recognized as being flexible and responsive to customers changing needs and new market opportunities. However, the small firm, especially one formed by a scientist or an engineer who is interested in exploiting a particular market niche entered by his expertise, is particularly vulnerable to both external conditions (over which the scientific entrepreneur has only very limited control) and internal constraints which result from limited resources. Many small firms, for example, experience practical problems of raising finance, finding premises, having insufficient manpower resources, acquiring market intelligence and developing marketing ability.

One of the ways of overcoming come of these problems and developing a competitive strategy is through collaboration with another firm, to gain "complementary assets" as Teece (ref 1) calls them. However, there are also dangers for the small firm in such situations and our research set out to investigate both these potential gains and losses.

The types of small firms which have been the focus of our research are increasingly recognized as important sources of innovation. Recent research, by for example, Pavitt, Robson and Townsend, (ref 2), has shown that innovation intensity amongst very small firms significantly increased over the period 1945-1983, and declined among medium sized ones. In electronics, instruments and electricals, small firms are contributing a major share of innovations. Technologically advanced small firms often carry out specialised functions which a principal company chooses not to perform itself but for which the subcontractor (or in this case partner) has special skills and equipment. However, the small firm is very often much more than a subcontractor and can be an equal partner in the development of new technology.

The spectrum of inter-firm collaboration for innovation ranges from pre-competitive R&D collaboration through to competitive R&D cooperation (ref 7). It includes activities which are co-ordinated by formal structures and other, more casual arrangements which have not reached the stage of any formal or legal agreement. The types of relationships range from planned, proactive strategies to those where partners reluctance to cooperate is overcome by pragmatism based on technological or commercial necessity. The propensity to collaborate with another firm is obviously dependent on the individual circumstances of firms, but is related, as we can confirm from the research, to such factors as the visionary approach of key individuals, the location of company control, sector and technological opportunities, competitive situation and the political/institutional framework in which firms operate.

Methodology

Over the last year, 27 case studies of collaborative partnerships were investigated. Five were collaborations between small firms, 16 were between small and large firms, and the remaining five were between large firms. All of the small firms in our sample employed fewer than 120 people. The sample firms were selected in three industrial sectors which we predicted would differ in the level and type of inter-firm collaboration. We chose biotechnology as a sector in which there would be "planned collaboration", electronics where collaboration is commonplace, and electrical/mechanical engineering where collaboration would not be the norm. For each case study, semi-structured interviews took place with both collaborating firms, covering the same issues. In this way different accounts of the same activities were gathered, enabling us to compare the perceptions, motives and assessments of the two partners. The analysis has primarily focused on the motivations driving collaborative ventures, the gains and losses and the issues involved in collaboration as a means of doing business.

Motivations

The fundamental motive for collaboration is to gain more efficient use of internal resources through access to external resources. Subsumed under this fundamental preoccupation, we have found a variety of different motivations for entering collaborative arrangements. These fall into four main categories: financial, technical, market and strategic which can be discussed in turn, followed in contrast, by a brief discussion of why firms may be reluctant to enter collaborative arrangements.

Financial Motives

For many small companies, the decision to go into a collaborative arrangement is taken to earn income and generate cash flow. Decisions are taken to collaborate as a means of earning money from providing technical input into a project based on the company's intellectual property. A company's motives may involve generating income from future sales of a product to a company with which it collaborates. Small firms can be in the position where collaboration guarantees a market from a large company, which has a vested interest in the product. In this case, companies trade expertise against a partner's need for new products. The incentive to collaborate is to gain better access to the purchasing requirements of the partner. By retaining the intellectual property rights (IPR) of the new products, the company can market the new products and sell to other customers. The motive is to build up a company's resources in order that it can have more control over future projects, where the preferred option is not to collaborate. In this instance the small firm collaborates with a user, a form of relationship much researched and discussed (refs 3 & 4). One of the case histories included in this paper is a good example of this tactic.

Often collaboration is a means of obtaining grants to support innovation. It is often easier for a small firm to obtain grants directly or indirectly through such programmes as Eureka and ESPRIT if it has a much larger partner. We know that in some countries collaboration is more actively encouraged than in the UK; and we also know that because of the paucity of UK Government funds, more UK companies apply for ESPRIT funds than do those of any other country, but overall only 1:20 applications are successful. The lack of support for innovation in the UK encourages firms to apply elsewhere for funding for collaborative projects. There is clearly a difference between the UK and other countries in the amount of support for both innovation generally and collaboration in particular (ref 8). This affects the competitive situation in which UK companies operate, and presents a further set of difficulties and options for both small and large firms.

Technical

A motive for collaborating which may have less immediate financial reward is where a firm identifies particular applications of a particular product, which if used in conjunction with another company's technology, will also benefit the partner by solving a technical problem. This may lead to a longer term collaborative arrangement, when markets are identified and developed. Here the collaboration has the effect of producing an extra dimension to the principal company's existing products. In other cases, firms collaborate in order to develop a solution to an internal problem but which can be applied to other areas, increasing production efficiency or generating income through marketing applications as adjuncts to original products.

Innovative firms of any size, even research centres of large corporations, may need to gain access to another company's technological resources. The firm or research unit which breaks new ground in technology often does not have the resources to exploit new developments because final product development and commercialization requires capacity which the initiating firm does not have. A firm which has complementary specialised knowledge in that field can take the innovation through the applied stages, developing a product which meets the needs of the user.

Small companies can develop very good working relationships with large companies which allow them to access laboratory and production resources. In one instance in our research (case no. 1), the motivation to collaborate with a particular large firm was to gain user expertise on site, as the larger firm would test, provide feedback and effectively be involved in the development of a new process. Another motivation for small firms to work together is sometimes reciprocal arrangements for evaluating each other's products/processes.

Markets

One of the problems small companies face is that their marketing and distribution resources do not match their technical abilities. Collaboration can be driven by the motives of gaining access to other companies' markets, and to other assets such as distribution networks, and personal contacts, by developing a joint product. Very small and very specialised companies may have the incentive to work with another firm because they are not large enough to employ people who can provide the necessary technical expertise to meet customer requirements as well as infrastructural support such as legal departments.

The opposite incentive to this is where small companies can provide an appropriate outlet for a foreign firm who wishes to enter the home market. Collaboration at the technical level may be essential to meet customer requirements in the UK, and UK technical standards.

Strategic Motives

For some firms, collaboration is only one tactic amongst a variety of operational practices. However, some small firms make a conscious management decision to pursue collaboration as a company strategy. Management applies that in order to develop from being a company which just develops existing technology in a novel way to gain a competitive advantage, it can collaborate, gaining technical experience and knowledge of markets in order to become a leader in technology. For instance, the UK biotechnology industry is very competitive and technologically dynamic and collaboration is successfully employed by both large and small firms (ref 9). Small biotechnology companies collaborating as a means of gaining income for their part in development is common practice. The incentive is for themselves to become large enough to accommodate the converging technological developments in pharmaceuticals and biotechnology.

Reluctance

While some companies overcome the problems encountered in collaboration, others either avoid it or have had experiences which make them reluctant to enter into another arrangement. Reluctance is based on a number of factors.

(1) The first is loss of control over the direction of the project.

A smaller firm may fear that a larger firm may dominate to the extent that the smaller partner is unable to influence the direction of the project and be unable to achieve its objectives. This may mean the dilution of profit. Loss of control may also mean loss of secrecy. One firm in our study was reluctant to collaborate because of an experience whereby a potential collaborator had imparted to a competitor confidential information given in good faith. Other firms are reluctant because of possible dependence on partners which may not be technically competent, again based on experience. Some larger firms are reluctant to enter situations where partners may contributing less but get more out of the collaboration than themselves. Others avoid collaboration with firms which are known to be predatory.

(2) The second is related to market structure. In a fiercely competitive market, collaborating with firms which may become competitors is not seen to be a good strategic move.

(3) A third set of factors is cultural or organizational. Some companies do not wish to enter into collaborations because of the "not invented here syndrome". This issue arose in all case studies which involved firms with headquarters overseas. Other firms do not

have the organizational flexibility to make or encourage informal arrangements which can often lead to successful collaboration.

Finally, some firms do not feel the need as they believe their own technical capacity generates sufficient innovation. From our case experiences, this arrogance did not serve them well.

Outcomes

We now turn to the outcomes of collaborations; the gains and losses. While the gains can be substantial, there can be serious problems for firms when things go wrong. Discussion on these gains and losses are categorized in the same way as the preceding discussion.

Financial Gains

A major gain can be increased turnover and profitability This has sometimes been critical in the survival of smaller firms as in the example of our case no. 1. The improved financial situation in the smaller firm in this case and in others in our study enabled other internal projects to be funded, and gave firms more choice in the direction they wished to take. In addition, some firms have been able to issue licences to their partner and so collect royalties.

Technical Gains

The first gain is the development of new products or processes which could not have been developed without complementary technical skills. Some of the development strategies involved formal, long-term arrangements while others were more casual, short-term interactions. We found that in most cases, even in the mechanical engineering collaborations, projects last around two years. In addition, in the majority of cases, firms continue to work with the partner on new ventures after the completion of existing projects. A potential outcome is the establishment of a new joint company to exploit the results of collaboration. However, we found only one instance of that occurring.

A related gain is that the number of contacts which can be used to access technical "know-how" can be extended through involvement in another company's network of contacts. Such as extension to one's own network of contacts proved very useful to firms in solving other technical problems, etc.

Market Gains

Collaboration can lead to improved scientific standing which can be exploited by the publication of results: either jointly or as an individual effort. The establishment of a track record, commercial and technical, in conjunction with an important principal company gives commercial credibility – this may reflect the principal company's scientific respectability or its commercial image. Both of these can be used as a marketing tool in attracting new customers. Scientific credibility may also be important in attracting the calibre of staff essential for maintaining the expertise level within the company. A commercial gain may arise when the partner is approached but decides that the work is not in that company's line and the customer is passed over to the partner.

Strategic Gains

Collaboration creates indebtedness and reciprocity – whereby a relationship is established which may be called upon when the needs arises in the future. It can result in a long term fruitful alliance, in that although the relationship may change and the balance of power and commitment may change, both sides benefit from a stable, predictable foundation. It can also improve the competitive position of both sides by the combination of expertise and resources. Both sides learn a certain amount from each other, such as R&D techniques, which can be applied to other internal activities. It can also widen the horizons of partners to other opportunities.

Strategic Losses

A loss may arise out of over-dependence on one major collaborator. This may be a problem where the company strategy changes and different priorities take over. In the case of one company we visited, dependence on a key individual who left the partner company, resulted in the relationship coming to a sudden end. A similar situation can also arise when a key individual is moved within the company.

Financial Losses

These can arise where the project does not develop as anticipate and where one side has not totally fulfilled its commercial commitments. We found one example where the process was too complicated for the users. The project did not bring anadequate return on investment and the amount of time and money invested by both sides has not been rewarded by adequate financial return. This could be put down as a failure of management to agree outcomes at the outset and/or a lack of monitoring.

Technical Losses

We found that some firms lost valuable technical advantages to competitors through the intentional and unintentional actions of other firms in revealing commercial secrets to other companies. Insufficient technical preparation on the part of one collaborator leading to product development delays can cause disappointment and frustration for the people working on the collaborative project.

Market Losses

Losses have arisen out of the length of time taken to establish a collaboration. This was because of the process of developing the right basis for a relationship meant that developments were delayed, and hence manpower and resources were tied up and a technical lead was lost. Failure to establish the ownership of IPR at the outset has meant, for some companies, that they lost out on a share of ownership of the new developments, and found themselves in dispute later.

Projects have been delayed – owing to resistance within a company at senior management level because collaboration runs counter to prevailing management style; because of the "not invented here syndrome" within international companies; and because the collaboration (and its success) has meant that other technical people or other groups within the principal company have not achieved the same results, and they are not willing to accept an imposed innovation. Collaboration therefore needs careful internal management and good communication within the company.

Lack of understanding of what was expected out of the collaboration has caused friction during relationships, but difficulties have been resolved by sharing anxieties. Differing priorities within companies has caused difficulties where the pressure to get a product to market on one side conflicts with a lack of urgency on the other. This has also been a problem where one side has decided to postpone the final stages of a collaboration which are of less interest to it but of prime importance to the partner.

Issues

The issues which we have been exploring have been identified from existing literature and by own experiences on this and previous research. Here we briefly raise a number of these issues. They may not only be the starting point for further discussion but we believe these are some of the more important points which firms themselves should consider when embarking on collaborations. We pose them as questions to indicate that there may be no definitive answers and also simply to challenge our readers.

1. Are structural changes occurring in industry at national and international levels which make it increasingly necessary for firms to collaborate in order to gain a "competitive edge"?
2. Related to the first question is the issue of "technological opportunities", i.e., is the conduct of collaboration affected by the type of technology and the stage of maturity of that technology? For example, in new industries such as biotechnology, is the nature of collaboration inevitably different from other sectors such as mature industries like electrical engineering?
3. Some questions about the establishment of collaborative ventures: is the conduct of a relationship affected by the way initial approaches were made? For example, would a relationship be different if a large company approached a small rather than the reverse situation? Are collaborations a result of "bottom up" or "top down" processes? Is there a danger in revealing too much to a potential partner? How important are "invisible colleges" and professional and information networks?
4. Expectations and realities: how much do firms need to know about their potential partners expectations and priorities? To what degree should firms articulated their own expectations? Who owns the Intellectual Property Rights? Who pays for what? Under what circumstances is secrecy an issue between partners? What constitutes failure: can financial failure at all be offset by learning experiences in technical and commercial areas?
5. Conduct of the collaboration: What length of time is optimal for collaborative ventures?; To what extent will corporate or political cultures affect the conduct of the relationship? To what extent will formalization or lack of formalization affect the conduct of the relationship?
6. Size and ownership: To what extent are these factors important in the nature of the relationship? Would there be any significant differences in the way a foreign multinational company behaved in collaboration compared to a UK-based company? Are large firms aware of the potential to be gained from collaborating with small firms? Are small firms aware of the power they can wield in a relationship with a large firm?

7. Communication: To what extent and at what level is communication between partners critical in transferring ideas and monitoring progress? Does professional trust have a role to play in collaboration? E.g., are there differences between the way scientists and production engineers interact within a collaborative project? How do relationships change as projects reach the commercialization phase?
8. Is collaboration a suitable method of doing business?

Case Histories

We now present three cases histories which illustrate some of the points raised in the above discussion. In each case, the identity of the firms has been protected by the use of fictitious names. Other cases have been published elsewhere (ref 5 & 6).

Case No. 1: Matthews and Muirs

The Actors

MATTHEWS is part of a worldwide electronics company. In the UK, the company has several semi-autonomous divisions. MATTHEWS is engaged in the design and manufacture of electronic equipment and employs about 800 people in this division. The market in which MATTHEWS competes is dominated by a few major players from Japan, the USA and Europe. It is important for MATTHEWS to be able to keep a high profile in the field, but as it does not have expertise in some areas of software development and electronics instrumentation, it is necessary to collaborate with small specialist companies in order to produce a range of equipment which can meet the applications requirements of customers.

MUIRS is a small electronics company, specializing in micro-processor controlled testing equipment for use with powders and liquids. It has progressed from being run as a "cottage industry" to a successful company, which although small, has the range of technical and marketing expertise which has ensured its high profile in its market niche. Venture capital was injected three years ago enabling the company to expand its operations, taking on a sales force, appointing dealers, etc. They now design and manufacture complex pieces of instrumentation which can be plugged into any system and customized to meet end user requirements.

Set up

The contact with MATTHEWS was made at an exhibition at which MUIRS had a stand. A marketing person from MATTHEWS, who was well known to MUIRS, recognized that MUIRS could solve a problem which was causing his company considerable difficulty, and so initiated the collaboration. A contract was drawn up soon after this which committed MUIRS to developing and supplying MATTHEWS with a item of equipment which would be central to a major new product. MATTHEWS was keen to establish a collaboration with MUIRS as a previous attempt to develop a system with another company had failed, and the company "were in some desperation". Within a short space of time after the deal had been signed, almost the entire senior management at MATTHEWS was replaced. The new management now felt threatened by the extent of the reliance on MUIRS, and felt that MUIRS had done rather well out of the financial arrangements. A whole new set of relationships had to be established, and which had to overcome suspicion on the part of MATTHEWS. Mr Muir went to Oxford to see the new MD at MATTHEWS to reassure them that MUIRS were not out to exploit them.

Procedure

The new management were to some extent pacified by what they were told by Mr Muir and agreed that there should be visits to MUIRS on a regular basis in order to monitor developments. In fact visits take place every two months, alternating between the two sites. MATTHEWS respects MUIRS technical competence, particularly in the software area, and its ability to structure the work load so that realistic deadlines are set and met. As both sides got to know each other, the professional trust which developed has improved the amount of technical support which MUIRS is able to give MATTHEWS. People from both companies spent a lot of time outside company hours getting to know each other on an informal basis.

MUIRS believed that MATTHEWS was in chaos, and that ithad been brought in to paper over the cracks in the design department. This department had been working on the project for three years without success, and MUIRS had been brought in as a last ditch effort to get things right. The previous effort had not addressed sufficiently the needs of the market.

Outcomes

MUIRS feel that its input has been critical in providing MATTHEWS with a product which, if not at the forefront of technology, will be sufficiently competitive for the company to maintain its market share.

On the financial side, MUIRS has gained a four year contract with a major international company, but the collaboration involved far more participation from them than had been anticipated, and hence the resources allocated at the start of the project were used up sooner than anticipated. Now MUIRS is giving free support to MATTHEWS outside the contract. MATTHEWS believed that it should have bought out MUIRS at the start, as it would have had more control over developments. It also believes that if it had not paid MUIRS so much money at the outset, then that company would have gone bust. However, other senior personnel felt that there was much to be gained from not absorbing small flexible companies into the large company structure. MUIRS has benefited from advice from MATTHEWS on how to develop its products so that customer problems would be reduced, and was taken to meet some of MATTHEWS's customers in order to understand operating difficulties.

MUIRS will gain financially from licence payments paid for the use of the software which it owns, and which MATTHEWS has no control over. It also gets sales for equipment as part of the product which MATTHEWS sells. It has also gained expertise and access to a extensive sales force. This is one instance where the smaller company's sales have increased more than the larger company's. The collaboration could also have been critical in the survival of the company.

MATTHEWS has gained a product which it would not have had otherwise. Its software people have learned techniques from MUIRS which can be used in subsequent projects. Without MUIRS, MATTHEWS would have lost market share, but it will take a long time before the money paid to MUIRS will be recouped.

The lessons to be drawn from this experience have been for MATTHEWS that collaboration with small companies can be very beneficial, if they are handled properly. It requires good management from senior levels downward to understand that small companies offer expertise and flexibility which large companies may lack. The success of a collaboration with small firms depends every much on the individuals in technical and marketing divisions getting to know their partners, developing professional trust and understanding their needs. For small firms, especially where they are undertaking OEM work, collaboration can give an insight into the operating needs of customers which is important in the design of equipment, and the provision of back-up support. It also shows that skilful negotiation can lead to advantageous financial and contractual terms, and that small companies do not always have to be disadvantaged in their dealings with large companies.

Case No.2: CIMB and LABEQUIP

The Actors

CIMB is one of the largest UK companies, with its 9 divisions operating in several fields, and its philosophy is to extend its operations into new areas of technology. This particular collaboration was based in one of its research centres which was set up to exploit and develop its recently acquired expertise in biotechnology. The research centre is organized into different sections which report up to a divisional head. Ideas for projects originate at senior scientist level and are evaluated technically and commercially in special business teams which contain scientific, financial and commercial professionals. Information and patent searches are undertaken to assess the value of the project in relation to work undertaken elsewhere.

LABEQUIP is a small company, employing 8 people, which designs and manufactures laboratory equipment, sometimes under their own name, other times under the brand name of another company. The company has particular expertise in the design and manufacture of new biological equipment for which it employs a molecular biologist as a product designer. The company's immediate goal is to increase profitability, and is investing heavily in design and R & D to give itself a portfolio of products with which to attack several markets.

Set Up

The collaboration came about as a result of an internal need for some equipment to be used for research in CIMB's laboratory. The assay process involved in the particular research project is labour intensive, and the project leader was looking for a means of automating the process. As the company does not manufacture instrumentation, an outside contractor was needed. The project leader actively looked for collaborators to help develop a new piece of kit. Coincidently, he was approached by LABEQUIP who were trying to sell him some of their new products. He had previously bought other equipment from them. LABEQUIP knew the area of interest, and were keen to develop a new product as a means of extending their own product range.

Procedure

About 3/4 months after talks had begun, a confidentiality agreement was signed when it was realized that the relationship could develop into one which had significant commercial potential. This set out what LABEQUIP could do in terms of exploitation of technology and sale of equipment. The development, however, had a low priority for CIMB in that the present assay could be done mechanically and the research was not much delayed by the lack of automation. The advantages to CIMB were more long-term in that future research would be facilitated by automation. Conflict arose because LABEQUIP had committed themselves to this project and were interested in short term solutions which would present commercial opportunities.

After six months development, the machine was working, but after further modifications, it failed to work. In their efforts to develop the product, LABEQUIP were not always careful in the preparation of the prototype equipment which was brought to CIMB for testing. When the machine failed to work, CIMB people working on the project became less motivated to develop their side of the technology. However, the good relationship between the project leader and LABEQUIP has been maintained and LABEQUIP personnel have continued to have access to the research laboratory, which would possibly not have occurred in arrangements with other companies.

Outcomes

A machine is being developed which utilizes CIMB's technology while LABEQUIP provides the manufacturing capability and knowledge of other potential customers. The machine will bring in some royalties to CIMB, and will be another advanced product for LABEQUIP. CIMB will gain a piece of equipment that will enable a more efficient use of resources in research, and by being involved in the development, gain an advantage over their competitors through first use of a significant innovation. The product will also have useful applications in other divisions of CIMB.

In addition, the project leader has gained a feel for the possibilities of other spin-offs in research in the laboratory. He has also learned about business priorities, in terms of both resource priorities in his own lab and about commercial priorities in other companies. He has realized that timescales in industry and in research laboratories are different. He has also learned how to overcome the problems faced by the difficulties in the collaboration – the failure of the equipment – through re-motivating staff. The project at the moment is a "heroic failure", but is likely to be a commercial success in the near future.

LABEQUIP have gained from being able to use the resources at CIMB, which have significantly contributed to the way in which the machine has developed and in the time taken to prepare a commercial product. They also have a potential market inside CIMB. Furthermore, there are other possible research areas in which the two sides could collaborate.

The case clearly illustrates the importance of understanding the priorities and expectations of the other partner when setting up the collaborative arrangement. The poor performance of one partner, in this case by the failure of prototype equipment, can cause a loss of confidence within the other organization, creating internal problems, and in the long run, a reluctance to enter similar arrangements in the future.

Case No.3: Electroica and TL Ltd

The Actors

Electroica is a small electronics company which has grown as a result of its scientific and engineering expertise in a rapidly changing and technologically advanced market. The company has many contacts through scientific and engineering networks which often lead to new customers. Collaboration is not unusual for this company as it has been involved in ESPRIT projects.

TL Ltd is another small company, this time with a more scientific base. It too has an extensive network of contacts. It was as a result of these contacts that when the need came for a collaborative partner to become involved in a project which was being funded by a third party, Electroica was approached to do some development work.

Set up

TL approached Electroica on behalf of the third party in order to develop a piece of equipment which had been initiated at TL but which required the input of electronics technology to make it into a genuine innovation. It was felt at the outset by both sides that the collaboration would be successful because the collaborators had a similar approach and a compatible level of expertise. The third party undertook to contact potential end users and conduct other market research.

Procedure

A consortium was formed to exploit the new technology and Electroica was encouraged to commit resources to the project. The technical end went well, the two sides were in regular telephone contact and visited each other a number of times over the two year life of the project. The people involved got on well on a personal level, and "professional trust" existed on both sides.

However, the commercial end was a disaster. There was little initial success in identifying where market potential lay and there were no firms sufficiently interested to stay with the project in order to have first use of the new technology. It eventually became clear that the wrong kind of market had been targeted, because alternative technology was already doing the job effectively – and at a lower cost than the system being developed.

At this time Electroica decided that they should clarify the position over IPR on the grounds that they had been led to believe that they were entitled to a share. However, it emerged that the patent rights belong to the third party. TL Ltd were involved in a specific piece of research as a commercial venture paid for by the third party which had control over the project via patents. Electroica found itself in the position of being too far into the collaboration and of having devoted too much effort to pull out and realized belatedly that it ought to have sorted out the IPR issue at the outset.

Outcomes

Finally a major British company became interested and had an application in mind. It was instantly obvious that this was the solution to the problem of application. This is now the basis of the present collaboration. The third party eventually paid Electroica for the work it had carried out, but it took a very long time for things to be straightened out. A lot of animosity was felt between Electroica and the third party, and TL Ltd felt that it was also being blamed for the way that Electroica was treated.

Several lessons have been learned. Firstly Electroica has learned that it is essential to be absolutely clear at the outset of who owns what. It has also learned to be cynical of market surveys which can have the effect of distorting development effort. It also now knows that collaboration takes a considerable amount of management time and that it has to be more selective in whom it collaborates with. On the plus side it knows that its organizational structure and general approach makes it possible for it to work in any collaboration. This is also true of TL Ltd who recognizes the need to be flexible in its work style.

Conclusions

A much more detailed analysis of our research will be published shortly in book form (ref 6), so we shall confine ourselves here to some comments about small firms' role in collaboration and, in particular, the role of intermediaries.

We found that small companies can gain considerable advantages from collaboration, not the least of which is having their technical ideas underwritten by larger companies. They can also gain by working with other small companies. Smaller companies do not have to be disadvantaged by their dealings with larger companies which often depend on smaller ones for technical leadership. However, we found that some larger companies do not realize the strategic value of supporting small firms through the sensitive handling of contracts, and technical relationships. Small firms have to find ways of making sure that they are able to defend their interests in the face of adverse behaviour by large firms. In both these circumstances, the role of intermediaries could be significant and beneficial.

The Role of Intermediaries

In our study we have found a number of instances of involvement by third parties in collaborative arrangements. Many of these were collaborations funded by European Community schemes, usually ESPRIT. Many involved university partners, which were seen by the firms as providing essential technical input. In three cases firms collaborated with ALVEY funding, and one within EUREKA. The only other instance where a third party, was involved in the setting up or funding of a collaboration, Case No. 3, was where a major insurance company, acting as an investor in new technology, paid for the development of a new product.

The problems faced by firms in the ESPRIT, ALVEY and EUREKA projects, which mainly concerned timescales and commitment, were more to do with the nature of industrial partners than with the programmes. Projects had been held up or abandoned in one case because the intended collaborator had not allocated equivalent resources, in terms of manpower and money. In another case, a change of policy caused the firm to pull out of that area of technology. However, the third party in Case No. 3 must share some of the blame for the problems which arose.

As many of the firms did have problems in collaboration, especially in the area of IPR, it suggests that there is a role to play by specialised intermediaries – such as lawyers, patent agents, accountants and management consultants. These are variously interested in the setting up, procedures and outcomes of collaborations. Their role would certainly involve the following major tasks:

- to challenge the assumptions and unarticulated expectations on which collaboration is being based;
- to ensure effective communication between the two parties. To do this well implies that the intermediary would be knowledgeable about the histories, personalities, corporate cultures and management styles of the two parties;
- to ensure both partners are well aware of the financial and legal consequences of the collaboration, particularly on the issue of intellectual property rights.

However, the potential return for lawyers and consultants acting as intermediaries on small firm business is limited which suggests that state agencies, or non-profit organisations (such as The Oxford Trust in Oxfordshire, set up to encourage entrepreneurship in the area), could play an important role.

One recent development in the UK, which may be important in the future, is the creation by the Government in 1987,of 13 Regional Technology Centres (RTC's). Their services are obtained on payment of an annual fee and the objectives of RTC's which are relevant to collaboration are:

- to increase awareness of available technologies and their benefits;
- to actively promote introductions between providers and users of technologies;
- and to co-ordinate and develop the flow of technology and skills between education institutions and local businesses.

However, most of these RTC's are only beginning to become operational. It will take some time before their effectiveness as marriage brokers becomes evident. It seems unlikely, at this stage, that they will be able to address the needs of firms identified in this study. There is a general issue of how much third parties are able to do as counsellors, and how much their intervention overcomes constraints rather than imposes further constraints through bureaucracy.

That the funding agencies were the most significant third parties in our study reflects the lack of a coordinated UK technology policy. It is reasonable to suspect that many UK firms get involved with EC schemes such as ESPRIT only because the absence of national support for innovation. Moreover, as such a small proportion of small firms receive direct research funding from government sources, far more could be done by the UK government, to make industry more aware at least of the benefits of collaboration and perhaps even to institute some mechanism for coordinating collaboration needs. Ultimately, however, it is the firms themselves that must act first and unfortunately some sections of industry are not sufficiently foresighted to recognize the opportunities which inter-firm collaboration could present.

Future research

Having studied the process of interfirm collaboration in the UK, the obvious next stage is to investigate whether the same issues, constraints, and procedures occur when the collaboration takes place across national boundaries. What would be a more opportune time than now to investigate cases of collaboration between European firms?

A similar research methodology would be pursued with case studies being conducted between pairs of collaborating firms, the only difference being that each firm within each pair would be located within different European countries. We are proposing to investigate such pairings between the following countries; - Spain, France, Germany, Belgium, UK and Ireland. Depending on the links established with other European researchers, and on time and funding, it is hoped that at least 10 paired interviews will be completed with each combination (i.e., 6 UK firms with 6 Spanish firms, 6 German firms with 6 French firms, etc).

As in our previous UK study, this research will:

- compile and analyze data on all aspects of the collaboration, including its establishment, its operation and its outcomes;
- undertake comparative analyses of the pairings using cross-evaluation questionnaires;
- identify critical success and failure factors;
- outline the roles of third parties;
- organize workshop/conferences for participating firms and sponsors;
- produce final reports, articles, etc, for dissemination to practitioners, sponsors and academic audiences.

It may be necessary to divide the research into stages, possible conducting an initial pilot study involving, say, only a limited number of UK/other country combinations before embarking on further stages which progressively involve more countries in combination with those just investigated. In this way, a comprehensive analysis of interfirm collaboration between European firms could be achieved.

References

[1] D.J. Teece, "Profiting from Technological Innovation: Implications for Integration, Collaboration, Licensing and Public Policy", *Research Policy*, 15, 1986.

[2] K. Pavitt et al., "Technological Accumulation, Diversification and Organization in UK Companies, 1945-1983", *Management Science*, 35 (1), 1989.

[3] E. Von Hippel, "Users as Innovators", *Technology Review*, 80(3),1978.

[4] B. Shaw (1985), "The Role of the Interaction between the Manufacturer and the User in Medical Equipment Innovation", *R & D Management*, 15(4).
[5] K. Dickson, H. Lawton Smith & S. Smith, "Interfirm Collaboration and Innovation; Strategic Alliances or Reluctant Partnerships?" in R. Mansfield (Ed), *Frontiers of Management: Research and Practice*, Routledge, London, 1989.
[6] K. Dickson, H. Lawton Smith & S. Smith, *Research Collaboration in Industry*, Frances Pinter, London, 1990.
[7] R. Rothwell & M. Beesley, "Patterns of External Linkages of Innovative Small and Medium-sized Firms in the UK", *Piccola Impressa/Small Business*, 2, 1988.
[8] M. Sharp, "Corporate Strategies and Collaboration – The Case of ESPRIT and European Electronics" in M. Dodgson(Ed), *Technology Strategy and the Firm*, Longman, London, 1989.
[9] G. Fairclough, "Strategic Partnerships in Practice" in *Proceedings of Biotech '88*, On-line Publications, London, 1988.

A Study of the Activity of Private Consultants in Technology Transfer

E. Bourgeois

Introduction

In the months August to October 1989, the Department which I head at ANVAR, Services aux Enterprises, conducted a study on the role of private technology transfer intermediaries in France during the period 1986 - 1988 inclusive.

Before describing the main results of this study, I will remind you (or inform you) briefly about ANVARs' activities and status.

The meaning of ANVAR, Agency Nationale de Valorisation de la Recherche, hints at its "historic" role: transferring research results from public laboratories to the private economy, i.e. industry.

A second role emerged in 1979: helping firms develop their own innovation projects by granting them financial support, usually 50% of project costs.

At total of 11,000 firms have now been aided in two main areas:

- development of a new (or improved) product or process,
- fees for consultancy or subcontracting of studies, in the course of an innovation project.

This result has been acheived thanks to the regional structure of ANVAR, which is organized into 24 regional branches, with autonomous power to decide on financial support for inovation programmes of up to 3 million francs.

As we do not only provide finance, but also advice on and financial support for consultancy and subcontracting, we have come to know a certain number of consultants, and have made a directory of them.

In addition, we are now considering the effects of ANVAR's new orientation on its traditional activity of "valorization", i.e. transfer of research results to industry. A component of this new orientation is the approach of considering the private consultants as possible partners capable of having their fees financed by our grants, even if they work for public labs, in the field of valorization.

Aims of the Study

- To evaluate the activity of consultants described in our directory in the field of technology brokerage.
- To estimate the respective importance of different customer groups for theconsultants: public labs, big firms, SME's, private inventors.
- To describe the market for this brokerage using criteria like : type of missions entrusted to the consultants, successes and failure rates, factors influencing success etc.

Method

The population under surveyed was the consultants described in the ANVAR directory and claiming the following activities : partner search, negotiation under mandate, opportunity search, licence trading.

The questionnaire was sent to 170 consultants, 55 of whom sent an answer, 36 of which were useful answers.

Results

Types of customers/missions

Analysing the total of 191 clients and 208 contracts (i.e. missions entrusted to the consultants), one can observe:

Types of customers

SMEs : 115
Big firms : 34
Private inventors : 30
Public labs :12
The SME clients are thus the predominant group. They more than doubled between 1986 and 1988, from 25 to 60 percent
On the other hand, public labs seldom refer to technology transfer.

Types of missions

Patent licence offer : 94
Competence and knowledge offer : 51
Competence and knowledge demand : 47
Patent licence demand : 16
Offers as a whole appear to predominate. For "competence and knowledge", offer and demand are equivalent.

What is more puzzling, is the fact that most of the offers of patent licences originate in the SME's, while big firms offer few patent licences.

This is in perfect contradiction with the belief that big firms are in a position of optimizing their technology portfolios through granting licences, while SME's should be in a position of "licensing in".

Industrial branches

The "high-tech" sectors only just outweigh the traditional ones : 52 cases (electronics, bio-medical, information processing) against 41 (agro-food industry, wood-processing, mechanical/machine tools).

Geographical location

There were 125 non-national (non-French) clients involved in the 208 cases, 84 non-European clients and 41 from other European countries.

Dynamic Aspect of this Market

During the 3 years under survey, the market expanded greatly; This growth is mainly related to "competence and knowledge", offers and demands.

By way of contrast, licence offers seem to be a "mature" product.

Successes and Failures

Clients

The highest success rate is for big firms, followed by SME's.

Missions

Those involving demand for competence and knowledge experience have the highest success rate. Offers, whether for licence or competence, have the highest failure rate.

These results therefore seem to be favourable to the "demand pull" approach to technology transfer (firm-level diagnosis, branch studies and so on ..), and against "technology push" (insertion into data banks and so on ...). This must be seen in context given that using a consultant yields a certain success rate, even if not as high as desired.

Success Factors and Difficulties

Surprisingly, "human factors" rank first in influencing success: they are seen as 75% of the answer, strongly ahead of the purely technological factors.

In terms of difficulties, they too appear not to be primarily technological, but financial, with money coming before the factor "insufficiency of the technology brought in".

Licensing Agreements in the Portuguese Manufacturing Industry

V. Corado Simoes

International technology flows have strongly increased in recent years. The ongoing process of globalization of firms' strategies and the growing intertwining of economic relations among countries simultaneously require and promote an acceleration of international technology transfer.

For intermediate countries, like Portugal, technology imports are a must to keep and enhance international competitiveness. The

portuguese scientific and technological system, despite some achievements in the recent past, is weak and unbalanced. Gross domestic expenditure on R&D corresponds to 0.4% of GDP and only 30% of it is undertaken by industry.

Licensing agreements may therefore be a relevant instrument to supplement the low capacity of firms (and the S & T system in general) to generate new technologies. This paper will briefly review the role of licensing in the process of technology acquisition by Portuguese manufacturing firms.

The analysis of licensing-in agreements in manufacturing industry during the period 1978-1985 leads us to stress the following:

- the strong concentration of sources of technology – five countries (France, Germany, Switzerland, United States and United Kingdom) accounted for about two thirds of total contracts. The importance of our EC partners as technology suppliers has to be acknowledged (62% of contracts) and remained stable throughout the period;
- the strong sectoral concentration – almost 70% of all agreements were addressed to two main groups of industries: chemicals and pharmaceuticals; and metal products, machinery and transportation equipment. The number of contracts in chemicals and pharmaceuticals shows however a declining trend in the 1980s.
- the relevance of foreign-owned firms – licensing-in is far from being the preserve of Portuguese owned companies. Foreign affiliates held a large share of contracts (around 40%), with intra-group agreements (i.e. agreements between firms belonging to the same multinational concern) being very frequent.
- know-how and trademark licences are the most common, while patent licensing is relatively scarce (occurring in only 13% of con-

tracts as against 76% and 63% for know-how and trademarks respectively). This feature is most probably related with the low technological level of Portuguese firms.

- the total number of agreements is declining – after peaking in 1981 (to a large extent due to the renewal of old agreements required by the Foreign Investment Institute) the recourse to licensing-in appears to have steadily decreased since then.

The above approach needs, however, to be supplemented by data on licensing payments, since the level of occurrence of contracts does not tell us about their real economic importance.

The main features of payments by manufacturing industry may be summarized as follows:

- Strong concentration in the destination of payments, four countries (United States, France, Germany and Switzerland) accounting for about two thirds of expenditures; the United Kingdom and the Netherlands also have significant shares. Reference should also be made to the major share going to the EC as a whole (above 60%).
- strong sectoral concentration, even sharper than for the number of agreements. Metal products, machinery and transportation equipment were responsible for more than 40% of payments in the period 1979-1985, and the share of chemicals and pharmaceuticals reached 37%. There two sectors showed opposite trends, however; while the weight of the former increased steadily over time, chemical and pharmaceuticals lost ground.
- strong concentration by type of firms: around four-fifths of technology payments were made by foreign-owned firms. Although not unique to Portugal this fact should not be disregarded, especially since the share of foreign affiliates exhibited a growing trend.
- technology payments increased steadily during the 1980's with the average annual growth rate reaching almost 30%. This increase was almost exclusively due to foreign-owned firms. Technology payments by Portuguese-owned firms in 1985 were, in current escudos, only 7% above the amount recorded in 1982.

In 1984, slightly over 7% of Portuguese manufacturing output was produced under license (8% if petroleum refining is excluded). The importance of licensing was, however, very different according to the sector, from 26% in machinery and transportation equipment to figures below 1% in wood and cork and the non metallic mineral industries.

On the other hand, the use of licensed technology was significantly different according to the ownership of firms: 30% of foreign affiliates output was based on licensing, while the corresponding figure for domestic firms was 1.2% only. The relative use of licensing by domestic firms appears to have decreased in the present decade, especially in chemicals and basic metallurgy.

The high propensity of foreign-owned firms to undertake licensing payments may be partly explained by international financial management and by tax evasion considerations. Anyway, it is also found in other OECD countries. More troubling appears to be the low and decreasing recourse of domestic firms to licensing-in, even though it is known that they import technology through channels other than licensing; the purchase of machinery and equipment seems to be the preferred vehicle for technological upgrading, especially where traditional industries and small and medium-sized firms are concerned.

Putting together all the information collected the most striking finding is that Portuguese-owned firms make little use of licensing-in as a technology acquisition device and, furthermore that the relevance of licensing is decreasing. This runs counter what might be expected, taking into account the low endogenous capacity to generate new technologies and the need for Portuguese firms to upgrade their technologies to face increased competition.

Licensing and other non-equity inter-firm arrangements can be very helpful for Portuguese firms in getting additional assets to enhance their competitive position. Much has to be done however to increase the awareness of Portuguese small and medium-sized firms about the advantages of licensing-in; such an effort must be co-ordinated with actions aimed at promoting interfirm cooperation within the European Community.

The acquisition of foreign technology through licensing and similar agreements must undertaken from a dynamic perspective. Technology acquisition must be seen as an investment – a prompter of further growth and innovation – and not just as a consumption good, enabling the firm to increase market shares or to penetrate in new markets. Today's spearhead technologies will be outdated techniques tomorrow. Portuguese firms need to pay increased attention to technology in their strategies and to use cooperative arrangements (both with other firms and with research institutions) to effectively upgrade their technological capacity.

Direct Acquisition of Technology by Irish Industry: Results of a Survey of Irish-Owned Companies Licensing Patented Technology

T.J. Allen, D.J. Cogan, D. O'Doherty and B.A. O'Sullivan

Introduction

The technology gap between Ireland and the more advanced economies within the European Community was highlighted in a recent Commission study (STRIDE 1988) and was shown to be significantly greater than the corresponding economic gap. The conclusion of the study was that intellectual and technological resources are the organic determinants of economic progress and that economic disparities can be effectively tackled only by reducing the technology gap between member states.

The closing of the technology gap is going to be difficult but it has been successfully achieved in the past both by countries and by individual enterprises. Mechanisms which have been used with varying success include:

- Internal R&D
  This is a long-term and expensive approach which requires patience, commitment and considerable intellectual resources. Small countries cannot hope to rely solely on indigenous research effort even where this is undertaken on a cooperative basis, either internally or externally.
- MNCs
  This is the most developed channel for the transfer of technology: it is estimated that two-thirds of all technology transferred across national frontiers moves between MNCs and their subsidiaries. However, when this transfer takes place to a subsidiary in a developing country (e.g. Ireland), very little of the technology is diffused into the local economy. A number of our research studies at the Science Policy Research Centre, University College Dublin (SPRC) have shown this very clearly.
- Capital-Embodied Technology
  This technology is acquired with the purchase of plant and equipment. It is a very valuable source of best practice technology but in an era of rapid technological change it is unlikely to give the recipient

a significant competitive advantage. The latter is more likely to come from proprietary technology rather than from technology which is already in the public domain.

- Mobility of Technological Manpower
  There are many examples of new technological enterprises and even new industries resulting from the migration of manpower with specific technical skills and know-how. Ireland's historical role, however, has regrettably been that of a net provider of skilled human capital to other countries (Onyenadum 1986).
- Direct Acquisition of Technology
  Technology is also bought and sold and exchanged on a reciprocal basis but its successful transfer and diffusion is associated with a complex interactive human learning process. The success of the transaction depends not only on the quality and appropriateness of the technology being transferred but also on the capacity of the recipient to absorb and to adapt it in an unfamiliar culture and environment. This is a very neglected area of Irish technology policy and is the subject on which this paper attempt to shed some light.

Direct Acquisition of Technology

The objective of the research on which this paper is based was to establish the extent and effectiveness of attempts by Irish industry to enhance its products and processes through one approach to the direct acquisition of technology, i.e., patent licensing. It also aimed to discover first-hand the experiences of managers who had engaged in technology acquisition and to make these experiences more widely known.

The focus of attention here is on indigenous industry because the transfer of technology from MNC parents to branch plants in Ireland, which accounts for much of the technology transfer into the country, is conducted privately through internal enterprise channels. The role of local management in the process is often passive and indigenous technical and entrepreneurial skills are deployed and developed only to a limited extent. To counteract this, policy makers offer inducements to foreign-owned companies to perform some R&D in Ireland in an attempt to give Irish scientists and engineers participation and experience in technological innovation. There is little prospect, however, that whatever the inducements MNCs will fundamentally alter the way they transact their operations worldwide.

In the case of indigenous industry the situation is very different. Small Irish companies must keep in touch with international developments in product and process technology in order to survive and compete internationally. Most Irish companies do not have the manpower or financial resources to engage in significant internal R&D and depend on acquiring and adapting technology from external sources. Data which have recently emerged, (Eolas 1988), appear to indicate that the resources explicitlya llocated to the acquisition of technology from outside Ireland are hopelessly inadequate. Chart 1 categorizes total expenditure on the acquisition of technology from abroad in 1986 under five headings.

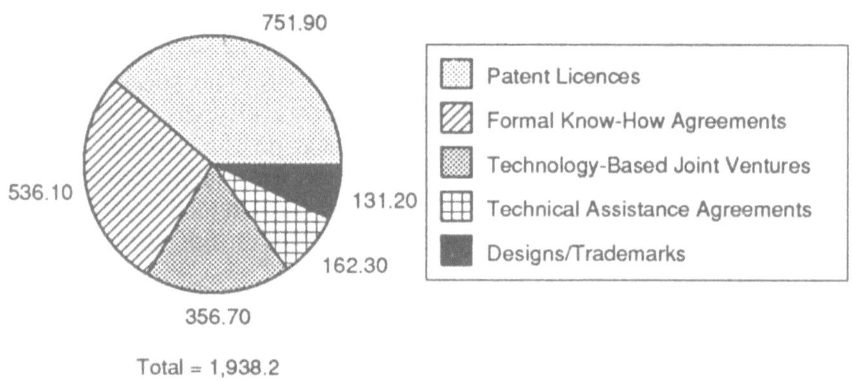

Chart 1: Expenditure by Irish Firms on Technology
from Outside Ireland IR'000, 1986

These are the first published data on the resources committed to the direct acquisition of technology from abroad by Irish-owned industry. There are grounds for speculating that there may be a measure of underdisclosure in the figures given: several respondents who indicated that they were party to a technology acquisition agreement failed to give any details of payments made. When due allowance is made for such underreporting, an expenditure of approximately IR£2 million compares very unfavourably with the resources being allocated to the purchase of technology by progressive industrial economies. To take one example: over 100 Dutch firms each report annual technology acquisition payments in access of IR£35,000 (Prakke 1988) while there is only one Irish-owned firm in this category.

Construction of Data Base

When this research study was originally conceived there were no data on the nature or extent of technology acquisition by Irish industry. A simple card survey was instituted and administered by the Irish Science and Technology Agency, Eolas, to establish the extent to which Irish industry, foreign-owned and indigenous, was active in the area. The companies were asked if they made or received payment (excluding payment to Patent Offices) either in Ireland or abroad for:

(i) patent licences
(ii) the use of established designs
(iii) the use of established trademarks
(iv) specific know-how agreements
(v) general technical assistance agreements.

Approximately 7,000 companies were surveyed and every effort was made to include companies, Irish and foreign-owned, who might be engaged in R&D or technology transfer. A total of 414 firms responded that they were engaged in technology transfer under one or more of the five headings listed.

Research Method

Sample Selection

It was decided to interview as many companies as possible in order to get a comprehensive picture of the current status of technology acquisition in Irish industry but the target population and the sample selection process had to be revised a number of times as the research proceeded.

1. Initially a random sample of 50 was made from the population of 414 companies who responded positively to the card survey. A letter was sent to each of these companies explaining the objectives of the research and requesting agreement to a telephone interview. After several interviews had been completed it became evident that many of the respondents to the card survey did not have a precise understanding of the term "technology acquisition". A number had manufacturing subcontract agreements and several had no technology agreement of any kind and no payment for technology had been negotiated or made.

2. In order to focus the research more clearly it was decided to interview only firms who had indicated in their response to the card survey that they had an agreement relating to patented technology. The objective was to identify all companies who had a

high grade and well defined technology acquisition agreement. A total of 123 companies (64 Irish and 59 foreign-owned) had responded that they were involved in a patent licence agreement.

3. After further consideration it was decided to confine the interviews to Irish-owned companies. The proposition has already been made that the technology acquisition process operates very differently in foreign-owned industry. From an industrial policy point of view it was considered more relevant to investigate thoroughly the pattern of behaviour within indigenous industry. All 64 Irish companies who had indicated that they had entered an agreement involving patented technology were written to and targeted for interview.
4. The population of 64 Irish companies thus contacted was found to comprise four distinct categories of technology agreement:
   (i) Purchase of technology from outside Ireland
   (ii) Purchase of technology from within Ireland
   (iii) Sale of technology within Ireland or abroad
   (iv) Intra-firm patent-holding companies (Section 34, Finance Act, 1973)

   For the purposes of the present study it was decided to interview only companies falling into the first of these four categories i.e. purchasers of patented technology from outside Ireland. This population comprised 28 companies.
5. The number of completed case histories was further reduced to 18 for the following reasons: four companies could not be contacted, four refused to be interviewed and there was lack of clarity in two cases.

Interview Method

The survey was conducted by means of an unstructured telephone interview which on average took about 20 minutes. The main objective was to get the respondents to speak frankly about their personal experiences in taking out a technology licence agreement so that the benefits and the problems inherent in such agreements might be more fully appreciated. A secondary objective was to obtain classifiable information about each agreement under the following headings:

(i) Subject matter
(ii) Sector of industry
(iii) Date of initiation
(iv) Country of origin

(v) How initial contact was made
(vi) Technical/marketing problems
(vii) Development of technology (if any) by purchaser

It did not prove possible to elicit information on all of these points from each respondent because the emphasis was on allowing the interviewee to speak freely on his own experiences.

Summary of Findings

Company Classification

Less than one in three of the 64 indigenous companies surveyed could positively be said to have had inward licensing experience but a number of interesting findings emerge from an objective analysis of their collective experience. The sectoral distribution of the 18 firms considered here corresponds well with the overall national distribution of indigenous manufacturing companies. There is no evidence of sectoral dominance on a proportional basis, with the possible exception of the Chemicals sector. Larger firms are however proportionally more involved than smaller companies in relation to the total population. (Table 1).

Sector	No. of Employees				
	U*	<50	50-99	100-200	>200
Food	-	1	-	1	1
Mechanical Eng	-	3	1	-	-
Electrical Eng	-	-	-	1	-
Metal	-	-	-	-	1
Non-Metallic Min	-	-	-	-	2
Building/Constr	1	-			1
Chemicals	1	1	-	-	-
Paper	1	-	-	-	-
Clothing/Footwear	-	1	-	-	-
Total	3	6	1	2	5

*Uncertain

Table 1: Licensing Companies by Firm Size and Sector (N=18)

Nature of Transferred Technology

In almost all cases, the technology that was transferred to Irish firms was either low-level or mature. Over 80% of identified agreements involve some element of process technology whereas 50% have a product involvement. Product and Process technologies are equally represented in agreements in the Food and Engineering sectors. All other sectors are distinctly biased towards process technology (Table 2).

Sector	Type of Agreement		
	Product	Product/Process	Process
Food	1	2	1
Mechanical Eng	1	1	2
Electrical Eng	1	-	-
Metal	-	-	1
Non-Metallic Min	-	-	2
Building/Constr	-	1	1
Chemicals	-	1	1
Paper	-	1	0
Clothing/Footwear	-	1	1

Table 2: License Agreements by Type and Sector (N=18)

Sources of Technology

Unfortunately there are little data available on licensors, apart from their nationality and the fact that almost all operate in the same industry sector as their licensee firms. The data in Table 3 illustrate the importance of EC countries as sources of technology.

Over two-thirds of the licensing agreements were concluded with EC countries, the UK being the major source. Food and Chemical sector companies are strongly influenced by the U.S.A., glass manufacturers look to W. Germany and firms in traditional sectors such as Building, Paper and Footwear have strong ties with the U.K. From Table 4 it can be seen that W. Germany and North America are the primary sources of process technology.

Nature and Duration of Agreements

Technology licence agreements vary widely in their terms and in their scope. The majority of agreements either explicitly or implicitly include

Sector	Nationality			
	UK	WGer	OtherEC	USA
Food	2	-	-	2
Mechanical Eng	1	-	2	1
Electrical Eng	-	-	1	-
Metal	-	-	-	1
Non-Metallic Min	-	2	-	-
Building/Constr	1	1	-	-
Chemicals	1	-	-	1
Paper	1	-	-	-
Clothing/Footwear	1	-	-	-
Total	7	3	3	5

Table 3: Sources of Technology by Licensor Nationality and Licensee Sector

Type	Nationality			
	UK	WGer	OtherEC	USA
Product	2	-	1	-
Product/Process	3	1	1	1
Process	2	2	1	4
Total	7	3	3	5

Table 4: Sources of Technology by Licensor Nationality and Type of Technology

a "know-how" element. Some exchange of technical know-how takes place, in many cases, long after the lifetime of the original licence agreement and extends beyond the bounds of the technical agreement. Long-term, beneficial relationships have been formed and sustained by this means. Table 5 shows that ten of the licensee companies have some form of know-how exchange arrangements currently in operation. Three of the remaining eight companies never had such an agreement and there is uncertainty about five firms.

Sector	Nationality UK	WGer	OtherEC	USA
Food	1	-	-	-
Mechanical Eng	-	-	2	1
Electrical Eng	-	-	-	-
Metal	-	-	-	-
Non-Metallic Min	-	1	-	-
Building/Constr	1	1	-	-
Chemicals	1	-	-	1
Paper	-	-	-	-
Clothing/Footwear	1	-	-	-

Table 5: Know-How Agreements by Source of Technology and Sector

EC companies appear to be more willing than non-EC companies to exchange technical knowledge on a long-term basis by way of agreement.

Type	Nationality UK	WGer	OtherEC	USA
Product	1	-	-	-
Product/Process	1	1	1	-
Process	2	1	1	2
Total	4	2	2	2

Table 6: Know-How Agreements by Nationality and Type

Table 6 shows that most know-how agreements are associated with process developments. Further insight may be gained by considering these agreements from the point of view of the size of the Irish licensee firm. License arrangements can be particularly beneficial for smaller firms who have limited technical resources. The relevant data are presented in Table 7. The data support the view that small companies are much more likely to avail of know-how exchange possibilities than larger ones.

Sector	No. of Employees				
	U*	<50	50-99	100-200	>200
Food	-	1	-	-	-
Mechanical Eng	-	3	-	-	-
Electrical Eng	-	-	-	-	-
Metal	-	-	-	-	-
Non-Metallic Min	-	-	-	-	1
Building/Constr	1	-	-	-	1
Chemicals	1	1	-	-	-
Paper	1	-	-	-	-
Clothing/Footwear	-	1	-	-	-
Total	2	6	0	0	2

*Uncertain

Table 7: Know-How Agreements by Firm Size and Sector (N=10)

An interesting question which remains to be discussed is the length of time for which agreements last. There are obvious difficulties in obtaining estimates of "average duration" since some agreements have only recently been put into effect, while the termination dates for others are uncertain. Another group of companies has terminated agreements prematurely, generally as a result of market failure. Table 8 presents a view (in some cases termination dates have been estimated).

Type	Duration (years)		
	<5	5-10	>10
Product	1	1	1
Product/Process	2	3	1
Process	3	2	4
Total	6	6	6

Table 8: Licensing Agreements by duration and Type (N=18)

One interpretation of these results is that agreements last, on average, between 5 and 10 years. This is probably valid overall but, since at least 3 of the agreements led to an unsuccessful outcome for the companies concerned, it would be preferable to assert that successful agreements typically last for 10 years or more. There is no evidence to suggest any significant relationship between sector and duration of agreement. Table 9, however, indicates that small firms tend to enter into long-term agreements with licensors whereas larger companies are slower to do so.

Duration	No. of Employees				
	U*	<50	50-99	100-200	>200
<5	1	-	1	1	2
5-10	1	2	-	1	1
>10	1	4	-	-	1

*Unknown

Table 9:Licensing Agreements by Firm Size and Duration(N=18)

Success and Failure

Not all the licensing agreements have a successful outcome but one of the most significant results to emerge from this study is that, in all of the cases considered here, the technology was transferred successfully. In general, small firms are weaker than larger firms with respect to the degree of in-house technical resources and associated supports for technology transfer and consequently could be more likely to experience failure. This is not borne out by the present study. The performance of the smaller companies in the study, once they had concluded agreements, is at least as good as that of the larger companies. It is evident however that many of the licensees had to put a great deal of effort into making the technology work and this is likely to have placed a greater burden on the smaller companies. The fact that all agreements were technologically successful is indicative of two things: first it reinforces the belief that relatively low-level technology was being transferred and, secondly, it highlights the importance of the transfer of know-how.

Successful transfer of technology notwithstanding, three agreements are known to have failed for market reasons and the outcome of another is uncertain at this time. All cases of reported failure are with product-related technology i.e. the product technology was not suited to the market for which it was intended. One third of product-related (with or without process involvement) technology agreements failed whereas none of the process-based agreements was considered a failure. Since process innovation is generally undertaken with greater market certainty that product innovation, this is not an unexpected result and reinforces the need for a sound understanding of the market prior to entering into a product licence agreement. Several of the failures were licences from North America, where the products had been very successful. The licensees assumed that the product would successfully transfer to Europe and failures resulted from misjudging the market.

The importance of transfer of know-how has already been highlighted. It has been suggested that continuing access to know-how is important, particularly for small firms, and Table 10 strongly points to this as a success factor. In all cases where arrangements to exchange know-how existed the outcome was successful. In two-thirds of the cases where there was no corresponding exchange the outcome was a commercial failure. The significance of know-how must therefore extend beyond the confines of technology to include a commercial dimension.

Agreement		Outcome	
	Failure	Success	Uncertain
No	2	1	-
Yes	-	9	1
Uncertain	1	4	-
Total	3	14	1

Table 10: Commercial Success of Failure Related to the Existence of a Know-How Agreement

Another factor which is strongly related to success is the movement of people. In the successful cases, employees were more likely to have been transferred for short periods in either or both directions. The more complex the technology and the more novel to the licensee, the more important this becomes. In some of the more successful cases employees of the licensee were trained by the licensor prior to implementing the technology.

Recommendations of Licensees

The following recommendations are taken directly from the individual interviews. Very little editing has been performed in an effort to preserve the recommendations offered in as near to their intended form as possible. Not all of the managers offered recommendations, but most made very clear and, in some instances, very strong ones. The authors believe that this simple compilation offers an exceptional amount of wisdom and is well worth the effort of perusal.

A. The basic advice offered, in relation to a licence for a food product, was "if you cannot write a preamble to (the basic terms of) an agreement in simple language on a half sheet of paper, do not go any further." It is vital for the licensee to have a market and not to rely purely on the licensor's appraisal of the potential market.

B. The basic lesson learnt by the manager interviewed, in relation to a licence for a consumer food product, was to be prudent in negotiating licence agreements: look for a low purchase price and go for performance-related royalty payments of say 7.5% to 10% of sales. It is important to travel and to share ideas with others but the cost of travel for a small company is relatively high.

C. This company which has long experience of licensing new seed varieties for agriculture, considers that the plant breeding business requires a long-term development commitment. This is primarily due to the long trial-period requirements associated with the development and testing of new varieties. Since the development of new seed varieties is expensive, it is essential to establish international, personal linkages, particularly in regard to countries within Europe with similar climatic and soil conditions, e.g. the UK, Germany, France, the Netherlands and Czechoslovakia. A build-up of mutual trust and understanding must be achieved before any exchange can be contemplated. The difficult part is making the initial contact. Potential transferrers tend to be wary because of ineffective patent protection and as a result relationships develop slowly. Reciprocal visits are very important not only to exchange information but also to develop trust. The Irish market is small and Irish seed producers have little incentive to incur the considerable risk of developing new varieties. Nevertheless foreign producers are willing to enter into agreements with suitable Irish producers and value any information on the performance of the new seed varieties.

D. The managing director interviewed in relation to a process licence for an animal feed product emphasised the vital importance of travel. It is crucial to build up a network of informal contacts in the industry. Industry associations should be used to arrange visits to Ireland and these visits should be reciprocated. The experiences and views of other licensees of the same technology should be solicited.

E. This company licensed a process for manufacturing insulating material. It consciously sought out the industry leader in the belief that such a firm would be using and/or developing leading edge process technology. The strategy of looking first at the quality/status of the potential licensor appears to have been successful. At the outset this company did not have a precise view of its technological needs. In this situation of know-how agreement was considered preferable to a patent licence as it provides broader access to technology and allows a firm to keep abreast of technological developments. Exchange of personnel is considered essential for the successful transfer of technology.

F. The manager interviewed with reference to a licence for a mechanical engineering product believes in a strategy of first seeking out sub-contracting opportunities and then buying designs from the potential licensor. The first step is to identify firms who are looking for sub-contractors or sub-assemblers. Trade fairs are an excellent way of building up valuable contacts. A sub-contractor can learn about the design practices of the licensor from the technical specifications provided. The next step is to identify customers who might be in the market for products with minor variations to those produced for the licensor. Finally, develop the local market and keep the product line up to date by purchasing designs from the licensor.

G. The manager here licensed a surface-coating process and believed that the indirect effects of his technology agreement far outweighed the direct ones. He believed that the experience had greatly widened the horizons of the firm, educated the staff and created many new opportunities. It opened up business opportunities with MNCs, who were more comfortable dealing with a company which had been exposed to US manufacturing technology and practice.

If the situation were to occur again the company would avail of the opportunity to update its technology on a wide front and explore further possibilities for technical cooperation and exchange. This manager also emphasised the need to allow for cultural factors when dealing with licensors in other countries. Considerable frustration can be caused for example by simply not appreciating the impact of different time zones and working hours.

H. In the case of a company with a licence relating to agricultural machinery, the manager was convinced that licensing was very satisfactory for expanding or upgrading a product line. A user in this industry is often well placed to recognise a need and propose a solution. However, there is need for caution because such "solutions" may not be viable. Foreign, and particularly continental designs may not suit Irish conditions. Travel is important in order to get ideas and to see what is being used abroad. Unfortunately there is no agricultural machinery association in Ireland but use can be made of the UK organisation. Much more should be done to stimulate licensing activity. The IDA brought over potential licensors from Germany, Canada and the USA but nothing appears to have come from this visit.

I. This company had a licence for a consumer electronics product and recommended that the total cost of licensing, which includes much more than the direct payments, should be carefully looked at. A significant commitment of internal resources will be necessary to make the transfer effective and produce a viable product. Attempts should be made to estimate the total potential cost of the project and this requires a detailed knowledge not only of the product and the technology but also of any changes or developments which will be required. The manager recommended that some form of local test market be carried out before getting too deeply into the agreement.
The company moved some of its manufacturing personnel to the licensor's plant for learning purposes. In retrospect they believe that this was very helpful and should have been done to a greater extent. They also believe that great care should be exercised in selecting people to be sent. Mistakes were made in sending people who were not sufficiently proficient technically and in not exploring the details of the technology more thoroughly.

J. The manager of a steel-making process advised similar firms to watch what the large companies were doing and then to come in on their coat-tails, applying their technology as quickly as possible provided the investment levels were not excessive.

Travel was also considered to be very important. Through the International Iron & Steel Institutes the Chief Executive makes initial contact and this is followed up with information contacts to get access to steel plants abroad in return for reciprocal visits.

K. The manager of a glass manufacturing company, referring to a licence for a proprietary method of drawing down and feeding molten glass to the work area, emphasised the importance of having internal technological competence and of keeping up to date through regular contacts with other glass manufacturers. People with relevant expertise are scarce and the company now employs an American who worked as a technological assessor for the UK Glass Institute.

L. This company, also in the glass industry, recommended trade fairs as an excellent source of contacts and ideas. In relation to a process agreement, the importance of seeing the process working before making an agreement was underlined. The workforce in the licensee firm needed plenty of time to get adjusted to the change. "Tell them about it, show it to them and train them to operate it before you introduce it." Make as much use of the licensor as possible, develop a close relationship and look for long-term back-up.

M. In the case of a product licence in the construction industry the manager emphasised the importance of "cultural factors". Markets in this industry are conservative and wary of new products. This arises from a tendency to specify products which have a proven record of performance and to ignore the functionality of the product itself. Knowledge of the market and its method of operation is critical. Specific details of who is going to sell the product and through what channels cannot be neglected. An appreciation of the time involved in setting up meetings and looking for the best applications for the product are also important. Develop a continuing open relationship with the licensor. In the manager's experience some companies are suspicious and reluctant to divulge information. Problems with the licensor must be sorted out quickly and frequent visits are essential. Travel is important for getting ideas and established contacts. As the company is large the financing of such trips is not a great problem.

N. The survival of this company in the construction sector forced it into the market for technology. A substitute material for asbestos which was discovered to be a health hazard was needed and urgent industry cooperation was imperative. The company discovered the immense contribution which its existing staff could make. It also learnt that travel is vital for success and that it must be the best people, preferably those with hands-on experience, who must be sent out to evaluate and learn.

O. This company licensed a new fine-chemical product. The manager concerned expressed the view that in an industry like fertiliser processing, where there are very few participants even in the UK, it is important to know them all well. Conferences were advocated as the best means of establishing contacts.

P. The manager here negotiated a chemical process licence and advocated the benefits of travel: "travel is the best investment you can make. If you can pick up just one item of information it is worth it." There are only two associations, with perhaps 60 members world-wide, and their annual meetings should not be missed. Reciprocal visits should also be arranged. Company personnel make one or two plant visits per year and have five or six visits from groups from abroad. When contemplating a plant purchase go to someone with a reputation not only for design but also for production. If they say they have 10 plants of the specified type in operation visit them all and talk to the operators.

Q. The manager interviewed about a licence for a new paper product expressed the opinion that, with any new product concept, there is a considerable learning curve particularly with the market. He considers it vital to be sure of a market before looking for the technology. The technology must be good and, if possible, fully developed. When dealing with a small donor company, the manager believes that ensuring long-term product/process technological support is more critical than obtaining the most advantageous financial terms. The exchange of information with other licensees of the same product is very important in that such contacts facilitate the identification of potential markets and the resolution of technical problems.

R. The manager interviewed about a process for colouring injection-moulded parts, believes that Irish companies should be able to get specialist advice on the negotiation of terms in licensing agreements. If he were to do it over again, instead of negotiating a flat percentage payment on sales, he would try to get agreement on a declining percentage over time. He would also try to limit the period of the payments to something less than the life of the patent. He has tried to find technology in this way again but has not been successful. Large firms are reluctant to cooperate with small firms, principally through fear of competition. He believes that the EC has made this worse and refers to the EC as a "nightmare", complaining that it has not only blocked potential inter-firm collaboration but has made it impossible to work under the old system of "gentlemen's agreements". He is now searching outside Europe for agreements. In his view, government help is needed to send managers abroad to search for ideas and to establish suitable alliances.

Conclusions/Recommendations

Conclusions

1. The patented technologies being licensed by Irish firms in the mid 1980s were, with one or two exceptions, relatively mature. The licensors were not providing a point of departure, in a technological sense, for the licensee company.
2. Cases are strongly concentrated in the traditional sectors and reflect fairly well the sectoral distribution of Irish industry. Highly represented are food and mechanical engineering. Building & construction and non-metallic minerals (glass) are also to the fore but electrical and electronic firms are not prominent.
3. The way in which contact was established appears to be random. It was frequently the result of an aggressive personality within the licensee company who took on the task of building an international network and searching for business-related opportunities. This required extensive overseas travel by somebody experienced and appropriate to the task.
4. Technology transfer is highly dependent upon the movement of people. When employees of the Irish firms were temporarily transferred to the licensor firms there was a much higher incidence of successful transfer. The quality of the employees engaged in the exchange of information was critical. A continuous, interactive relationship between licensor and licensee enhances the effectiveness and benefits of an agreement.

5. The standing and reputation of the licensor is as important as the quality of the technology on offer. Small licensors have particular difficulty in providing the continuing back-up and interaction which is necessary long after the initial agreement has been signed.
6. Informal networks represented a very important mechanism for transferring technology and for developing licence arrangements within an industry. These networks appear to develop at varying rates in different industries. In some industries they develop fairly rapidly, in others only after a considerable investment of time and numerous contacts. There appears to be an individual "culture" in each industry which determines this rate. The reasons for such large inter-industry cultural differences are not known.
7. The technology transfer process itself is conditioned by cultural factors. Unlike science, technology is not universal and cannot be successfully transferred without taking full account of the culture and environment of the firm and country (Allen 1984, Schumacher 1973, Pavitt 1983). The failure to recognise this characteristic of technology led to serious difficulties in several of the cases.
8. Evaluation of technology acquisition opportunities was done on an ad hoc basis with little indication of any rigorous "return on investment" analysis. The views of informal contacts within the industry and the reputation of the licensor were the factors which were most frequently mentioned.
9. Insufficient prior analysis led to initial failure in several cases. The licensee was then forced to invest internally in development work in an attempt to rescue the situation. In some cases this paradoxically resulted in remedying the initial failure and also in a changed attitude towards internal development and the building up of an internal technical capability.
10. The form of payment adopted was highly dependent on the degree of market uncertainty. The preferred strategy of Irish licensees where there was a relatively secure market was a single up-front payment or an outright purchase of patent rights. When the market situation was less certain royalty payments were preferred.
11. Irish firms have, in principle, been well placed to acquire technology from abroad because small firms in a small economically weak country were not perceived as an economic threat. Closer economic integration within the Single European

Market is already showing signs of adversely changing this perception. A similar trend has been observed in Portugal.

12. Respondents did not make any spontaneous reference to assistance received from state agencies or other intermediaries in their quest for technology acquisition agreements.
13. The interviews highlighted the dangers of venturing too far away from the firm's existing line of business (in terms of market or technology). Some of the "failures" were clearly attributable to this problem. The literature supports this finding (Meyer & Roberts 1986).

Policy Recommendations

1. Irish companies need to be educated to the fact that worthwhile product or process technology is unlikely to arrive in a convenient, ready-to-use condition. Significant adaptation and further development is almost universally required in order to mould the technology to the needs of a different culture and environment. This has implications for the level of technical skills which even the smallest firms must possess in-house, or more likely have access to from outside. Consultancy services in technology transfer should be available at local level in every region.
   An integrated intelligence network, channelling both business and technical information from national and international sources into one or more focal points in each region should also be implemented (Ghoshal & Kim 1986).
2. The stimulation of awareness of, and the development of competence in, product and process innovation should figure much more prominently in existing industrial development programmes, such as the IDA Company Development Programme and the Eolas Technology Audit Programme.
3. The high incidence of process technology licence agreements in the traditional sectors of industry manifests a strong market need for best practice manufacturing technology. The existing Eolas Advanced Manufacturing Technology Programme and perhaps some new technology demonstration projects should devote particular attention to increasing the awareness of companies of the part which licensing can play in updating process technologies.
4. The paucity of technology licensing activity in the high technology sectors of indigenous industry raises questions as to the

effectiveness of existing industrial policies in that area. In particular the possibility of utilising the proposed Programmes in Advanced Technology (PATs) in order to facilitate the identification and adaptation of suitable licensing opportunities, as well as the undertaking of specific projects on behalf of firms, should be considered.

5. It has been strongly indicated by this study that the acquisition of technology from outside Ireland is unlikely to happen without extensive foreign travel by potential licensees. It is recommended that a specific programme be developed to stimulate and support such travel. One suggestion is for carefully planned and coordinated group travel in well-focussed areas of interest similar to the Danish and Japanese models.
6. The fostering of trade fairs in Ireland should be given at least as much priority as the fostering of international conferences. This powerful and relatively inexpensive mechanism for technology diffusion is now sadly neglected. Ireland's strategic geographical location between the US and Europe provides an opportunity for the country to become a "hub" for the interchange of product and process information.
7. More reliable and concise statistics are required on the extent and the nature of technology acquisition by Irish industry if informed action is to be taken to remedy the defects which have been highlighted here. Based on the experience of this survey, proposals have been formulated and presented to Eolas relating to a full "technology transfer" section in the biennial R&D Survey.

References

European Commission. "Research & Technological Development in the Less Favoured Regions of the Community (STRIDE)", Final Report, 1987.

Onyenadum, Eze. Unpublished Ph.D. thesis, University College, Dublin: "Staff Mobility as a Mechanism for Technology Transfer"

Irish Research & Development 1988. Eolas, Dublin, 1989.

Parkke, Fritz, TNO Centre for Technology & Policy Studies, The Netherlands. A private communication. 7 December 1988.

Allen, Thomas J., Hyman, D.B. & Pinckney, D. "Transferring Technology to the Small Manufacturing Firm: A Study of Technology Transfer in three Countries", *Research Policy*, 12, 199-211

Schumacher, E.F. *Small is Beautiful*. Sphere Books Ltd, 1974.

Meyer, M.H. & Roberts, E.B. "New Product Strategy in Small High Technology Firms: A Pilot Study", MIT Sloan School of Management Working Paper, May 1984.
Cogan, D.J. & O'Sullivan, Brendan, in conjunction with O'Doherty, D. & Allesch, J. "Technology-Based Regional Development: A Conceptual Approach", Eolas, Dublin September 1989. Chapter 7 of this study makes the case for locally-based technology transfer services and describes the type of technical support which should be available to SMEs.
Pavitt, K. "Sectoral Patterns of Technical Change: towards a Theory and a Taxonomy" *Research Policy* Vol 13. 1984
Ghoshal, Summatra & Kim, Seok Ki. "Building Effective Intelligence Systems for Competitive Advantage", *Sloan Management Review*, Vol 28 No 1, Fall 1986.

Section 2: European Initiatives and SME Cooperation

This section gives an overview of the current European measures and programmes that have an impact on SME cooperation. It includes contributions from four representatives of the European Commission and one from the Secretary-General of the EUREKA Programme, which involves 19 European countries. The Commission papers cover the range of collaborative initiatives which the Community has developed as its distinctive mode of operation in technology and related areas. While they indicate a great deal of goodwill towards SMEs, the major area of expenditure on science and technology, the Framework Programme, still largely reflects the priorities of the major European firms (and countries). It is not clear that the situation and potential of small firms is actually fully taken into account, even in terms of the competitiveness criterion which is the primary objective of the Programme. While a number of specific initiatives have been discussed and some action taken within individual research programmes (and recognising that the technology transfer and "structural" programmes are aimed mainly at SMEs) there is no SME research programme as such. These issues are further examined and elaborated on by R. Rothwell in the policy discussion in the last section of this book.

At a wider level, despite the setting up of the SME Task Force in the early 1980s, later converted into the Enterprise Policy Directorate-General (DGXXIII), a fully integrated SME policy and attendant programmes across Directorates-General has not been developed. There is evidence, including that presented to the Conference on which this book is based by Chris Hull of the TII (Association for Technology Transfer, Information and Innovation, Luxembourg), that the most useful help to SMEs is a "total package" of assistance. This point is taken up in J. Mulcahy's paper here where he advocates adopting a "total systems" approach to assisting the SME at the regional level.

In the EUREKA case some serious efforts have been made to identify the "programme" and "structural" constraints on SME involvement. Paradoxically, in one sense, given that all participants in EUREKA are self-funded (or receive assistance from their national administrations) participation here is more directly relevant to SMEs because the activities tend to be "nearer the market". The "precompetitive" focus of the Community programmes on the other hand tends to make them less relevant to the perceived, often short-term, needs of SMEs.

Here it is also important to keep in mind the question raised by Prof. Robert Chabbal in his Chairman's introduction to the equivalent session of the Dublin Conference – "What is the point of R & D, no matter how worthy, if it is not applied and used and the results widely diffused. Cooperation between small firms themselves and with larger firms is of great importance in creating but most importantly in diffusing and applying research results."

In a later section of this book the editor also addresses some of the issues raised by the involvement of other actors in collaborative research programmes of the Community, particularly the universities, and the need to formulate appropriate national policies and responses to this growing aspect of the cooperation phenomenon.

On a very positive note we should remember when reading these contributions on more formal cooperation arrangements that clear indications are now emerging from firms' experience of these programmes that it is the informal contacts and networks which grow up around them that are their lasting achievements.[1]

The first paper in this section reviews the makeup and impacts of the 1st and 2nd Framework Programmes and gives a general outline of the 3rd programme recently approved by the Council of Ministers. It outlines a number of initiatives that have been taken, particularly within BRITE/EURAM (Industrial Technologies and New Materials) for which the author has direct responsibility, to improve the position and role of SMEs.

The second paper outlines the SPRINT Programme and gives some indications on VALUE, the emerging programme on commercialising Community R and D for which the author is responsible. Both programmes emphasise technology transfer and are mainly aimed at SMEs, but have very small budgets relative to that of the main Framework Programme for Research.

The third and fourth papers outline the mainly "structural"-type assistance offered by DGs XXIII and XVI which are intended to give small firms and regions the capacity to involve themselves in more explicit research and technology transfer programmes both naturally and at Community level. While these show a growing element of consultation and cooperation between DGs, an overall strategy still appears to be absent.

The fifth paper outlines the EUREKA experience and the actual and perceived roles of SMEs in that context.

1 E.g. A Preview of the Impacts of the 1st Framework Programme on Irish Science and Technology Policy :Fitzpatrick and Associates, to be published by the Commission in 1990

The Framework Programme – International Networks and Consortia in R and D

A. Garcia Arroyo

Framework Programme

Clear recognition is now given at the Community level to the crucial role science and technology must play in strengthening European industry and enabling it to become more internationally competitive. Furthermore, in making its contribution to these overall objectives, Community research and development policy must include the following aims:

- to encourage stronger integration through more extensive intra-Community cooperation, with public funds serving as a stimulus to this cooperation, and
- to improve the social and economic cohesion of the Community by reducing the technological imbalance between Member States.

It is against this background that the second Framework Programme covering the period 1987-1991 has been developed, setting objectives and priorities as well as the overall financial allocation and its breakdown between the different major activity areas.

The philosophy of the Framework Programme is, clearly, to undertake research which, for one reason or another, is more appropriately, more economically and more efficiently conducted at a Community level. One particularly important aspect of the present Programme (1987-91) is the field of pre-standardization research. Progress in this field, relating to the establishment of European norms and standards through Community research, is seen as an essential element in the process of completing the internal market by 1992.

The Community funding approved for the Framework Programme is some 5.4 million ECU and since Community aid represents only 50% of project costs, the total budget involved amounts to 10 million ECU. This is less than 3% of the total expenditure on R&D in the Community. However, considering that the Community programmes are confined essentially to pre-competitive R&D, the impact of these activities can be considerably higher. The ESPRIT Programme, for example, coordinates 30% of all pre-competitive R&D undertaken on information technology in the Community. The R&D funds for manufacturing activities are considerably less at about 4% of the Member States' effort.

The principal aim of the Framework Programme is to encourage the development of international competitiveness in European industry. It does this by promoting scientific research and technological development at Community Level, complementing activities in the Member States.

The Decision of the Council of Ministers adopting the Framework Programme recognized the contribution that small and medium enterprises can play in improving overall industrial competitiveness. The decision requires that particular attention should be paid to their needs with regard to access to information, participation in Community R&D and ability to exploit the results. The Framework Programme is implemented by a series of specific programmes which fall within one of 8 activities listed in Appendix 1 to this paper. Line 3, science and technology for the modernization of European industry, comes out as the third largest activity after Information Technologies (Line 2) and Energy (Line 5). The dominant activity in Line 3, the BRITE/EURAM Programme, was launched at the end of 1988 with the first call for proposals finishing in mid-May of 1989. We received some 645 applications which involved 3318 partners, of which 57% came from industry. Almost half the industrial partners were SMEs.

Networks and Consortia

The Commission essentially acts as a concerned host and focal point in order to reflect the needs of industry, for whose benefit the programme has been initiated.

During the preparation of a programme there is always considerable consultation. In addition to industrial studies and surveys to gather information there is consultation with industrial bodies such as trade associations operating at the European level. In the process of bringing views together the Commission is assisted by the Commission's industrial advisors – in particular IRDAC (Industrial Research and Development Advisory Committee).

Working alongside the Commission there are also the Committees, with representatives from each Member State which advise in programme management. Here, complementing the work of IRDAC's Working Parties, teams of experts from across the Community have helped identify the technical areas which the programme should include. I would like to single out Killian Halpin of EOLAS who was for several years a most effective Chairman of the Advisory Committee for Primary and Advanced Materials.

All this takes place when the programme is being prepared. We also network out into the Community during the running of the programme. Groups of experts assist in the definition of the Priority Themes which are included in the annual call for proposals. Similarly, when it comes to assessing the applications for support, a large team is required. This year some 140 industrialists and academics helped us select the proposals which should be supported.

You can see that before the research even starts we have established networks of experts who, working together, interact with one another to help break down established barriers.

The Commission has worked together with industry to help formulate its technology needs and thereby its research priorities. A recent example is with the clothing and textile industries where many problems have made it difficult for the industry to come together. After careful preparation, encouraging progress is being made towards establishing a network which would increase the awareness of R&D, technology transfer and training within the industry and be a source of consultation for the Commission.

Research consortia come together either on a project basis or for coordinated activities. The formation of research consortia for collaborative projects is the major activity. In the case of experienced partners existing contacts can be exploited. The task is much greater for those organisations which are new to research at the Community level and particularly SMEs. By nature of their size and limited resources SMEs can experience difficulties when participating in Community research. In particular, SMEs are not as easily informed as large companies and there may be difficulties with finding their share of project costs and dealing with languages.

The Commission's marriage broking service, operated through the matching of "expressions of interest", is an important instrument for finding partners. Within the BRITE/EURAM programme we have also introduced a means of helping individual SMEs to establish their credentials as potential partners. The Feasibility Awards, introduced earlier this year as a pilot scheme, fund SMEs for up to 6 months R&D in order to establish the feasibility of an innovative device, concept or process. In assessing applications for the awards, we are looking particularly for those projects which have scope for incorporation into the much larger collaborative BRITE/EURAM projects. We received nearly 600 applications demonstrating a very strong interest from SMEs. Sixty awards were made in 1989 and a similar number is expected in 1990. The success of this experiment will be measured by the increased awareness and participation of SMEs in research collaboration at the European level. Because of its SME responsibilities, DGXXIII is a co-sponsor of the Feasibility Awards.

In addition to the "expressions of interest" and "Feasibility Awards" special meetings such as the BRITE/EURAM "Technological Days" and "Proposers' Days" help SMEs find out about programmes and meet potential partners. The BRITE/EURAM Contact Persons in each Member State provide local advice which is particularly beneficial to SMEs. The "Euroguichets" supported by DGXXIII also help in informing SMEs.

Once the research starts, the focus must be on the consortia undertaking the research. The extent to which there is effective collaboration will depend on the planning agreed at the outset which has been confirmed in the contract negotiations with Commission Staff. It is at this stage that the benefits of collaborative research become clear:

- the ability of partners to participate in R&D activity many times larger than if it were conducted and funded in-house;
- access to complementary skills;
- opportunities for making contact across national and sectoral frontiers for building confidence and even leading to other forms of cooperation such as training, marketing or joint ventures;
- the participation of partners from different Member States can also lead to results which can be taken up in the process of standards making at the European level. This leads to a further breaking down of barriers which is necessary for realizing the Internal Market.

It is worth noting that there is already a substantial SME participation in Community funded R&D activities. The BRITE Evaluation Report concluded that participation in Community R&D programmes had already helped SMEs to become involved in transnational research, and establish linkages with large companies and universities. SMEs have also benefited from the resulting business and commercial contacts. In fact, in the earlier BRITE Programme 35% of industrial partners were SMEs and in budgetary terms 21% of the budget went to SMEs. A similar pattern is expected in the new BRITE/EURAM Programme.

There is also another R&D arrangement which is of major importance in making the most effective use of Community resources. I refer to the coordinated research which has been a characteristic of the COST actions but is also found within Community R&D activities. Here the role of the Community is not to fund research but the coordination of research in a particular area. Researchers, whose work may typically be partially funded from their own Member State, meet to plan their common goals. This coordinated approach is not only a valuable information route but a means of avoiding duplication, and of making the best use of resources such as by sharing equipment. A typical example is the CEAM activity in the field of magnetic materials. Here more than 70 laboratories, involving some 150 researchers, are working together. Almost half the participants are industrial including several SMEs. This activity is most effectively coordinated by Professor Michael Coey of Trinity College, Dublin. Coordinated activities are an important instrument where there are already developed and established research activities within the Community and for which real progress requires a critical mass greater than the individual research teams (High Temperature Super Conductivity is a recent addition to the range of coordinated activities supported by the Commission).

Scale of SME Involvement in BRITE/EURAM

The involvement of SMEs in the projects selected for support in the 1989 round of BRITE/EURAM is set out in the Appendix 2 to this paper. There are many observations to make from this detailed breakdown, but for our purposes the most important are:

- almost 60% of projects include an SME
- in 9% of projects all the industrial partners are SMEs
- one fifth of prime proposers are SMEs
- about half the SMEs involved have less than 50 employees
- Many SMEs are the sole representatives from their Member State in a particular project. This characteristic is more pronounced within smaller Member States.

From this we can conclude that small companies, including the very small with less than 50 employees, are playing a real part in BRITE/EURAM. It is also clear that there are a significant number of projects where SMEs are the sole participant from a particular region or Member State. In monitoring the progress of BRITE/EURAM we will need to examine whether this is too much of a burden on the SME, particularly when compared to situations when an SME is able to work locally in conjunction with another partner.

The Challenges

With the internal market in mind and only a limited financial resource, how should we manage the research programme to bring the maximum benefit for Europe? For example, should we be thinking of giving preference to proposals where there is the substantial participation of new partners?

More and more we can see the role of standards and codes of practice as competitive instruments in trade within the Internal Market. This is of particular importance to SMEs who have much to gain from workable standards in their role as customers or suppliers. The need to have appropriate standards in place in good time means that we must develop better links between R&D and the standards-making process. There are tasks here both for researchers and the standards-making bodies.

With regard to the interface between our research programmes and SMEs we recognize that there is always scope for improving access to information about our activities. We must also keep under review our apparently complex application and contracting procedures.

Turning to the measures of success, we are now at a time when a significant number of Community-funded R&D projects are coming to an end. Dissemination of the results must maximize the benefit of the Community's investment. The task is to balance the need for respecting the commercially sensitive value of the results with the need to optimize access to the benefits. Concern has also been expressed that dissemination of information on research should avoid revealing information on Europe's industrial R&D strategy to competitors.

The final challenge concerns those sections of industry, often made up predominantly of SMEs, which have no or a very low research base, either in-house or in a national research organization. Often the success of sectors – such as the tanning or the drop-forging industries – is a significant contributing factor to the competitive performance of their customers – tanning with footwear and drop forging with the motor car. In many Member States there is not even a research base on which SMEs can call. There would seem to be a need for cooperative research at the European level bringing together, for example, research expertise in one Member State with the technical requirements of SMEs in another.

IRDAC has been giving considerable attention to the need for cooperative research and the practicality of operating it at the European level. It has come up with a scheme (CRAFT) where groups of SMEs from different Member States find an appropriate research centre to carry out workfor them in which there is a common interest. At the time of writing the Committee is considering a recommendation that the Commission fund a number of pilot projects as a means of testing the concept.

Conclusion

At the end of 1989 there is now before the Council a proposal for a new Framework Programme for the period 1990-1994. It includes a number of characteristics which should help in addressing the challenges I have set out earlier.

- The new Framework Programme includes only 6 lines and it is intended that each should account for a single programme.
- There will be a single programme covering industrial and materials technologies including material recovery and recycling.
- There is a strong emphasis on pre-normative work.
- Priority will be given to large integrated projects.

The new Framework Programme offers an opportunity for a more strategic approach to R&D needs. We will be taking a more systematic look in conjunction with industry at R&D priorities and the type of actions within each programme by which these are best met.

More integrated programmes providing a single point of contact should help avoid some of the confusion which can arise from a large number of separate programmes. SMEs, in particular, should find it easier to establish how EC R&D programmes might be of interest.

With regard to cooperative research aimed at SMEs, we will also be examining how such an activity could be included within the programme.

The 3rd Framework Programme is an important step towards a more integrated and effective industrial research strategy for Europe. It is essential that SMEs from all Member States are able to play an effective role in both conducting the research and exploiting the results.

Appendix 1

Framework Programme 1987 - 1991

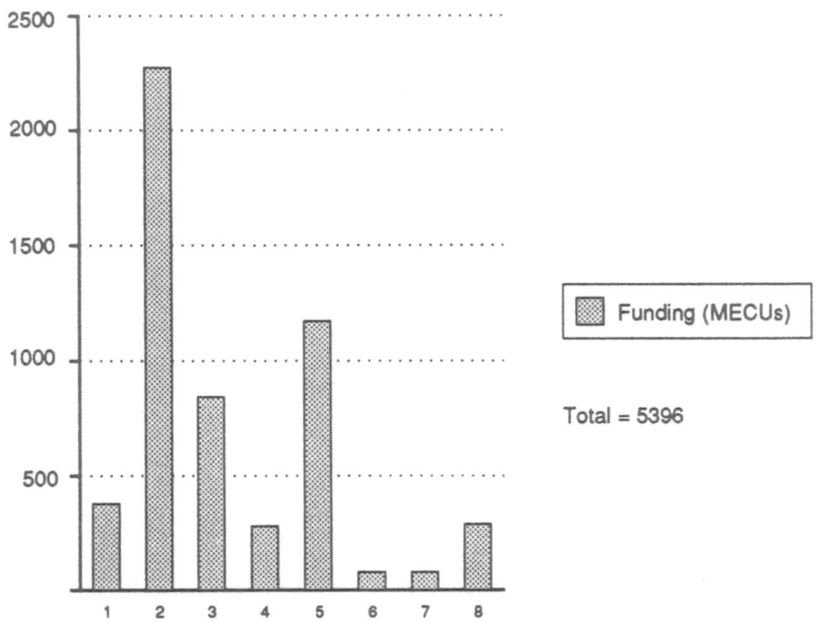

1. Quality of life
2. Towards a large market and an information and communication society
3. Modernization of industrial sectors
4. Exploitation and optimum use of biological resources
5. Energy
6. Science and Technology for development
7. Exploitation of the sea bed and use of marine resources
8. Improvement of European S/T cooperation

Appendix 2

BRITE/EURAM Programme 1989

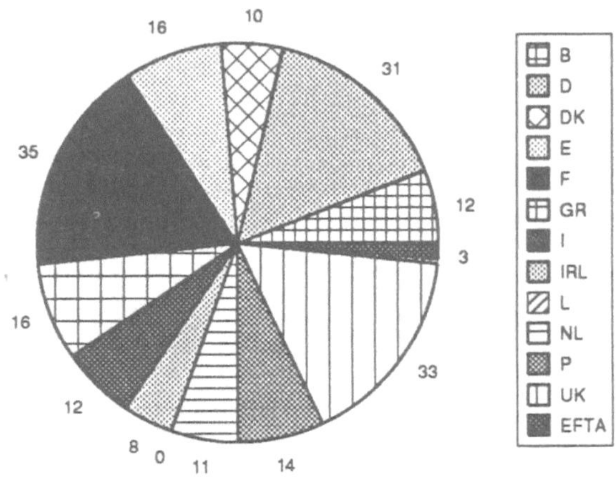

1. No of SMEs by member state

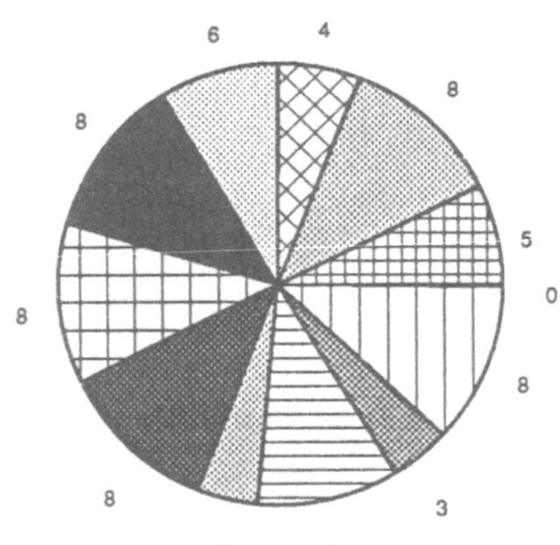

2. Projects where SME is on its own in a single member state

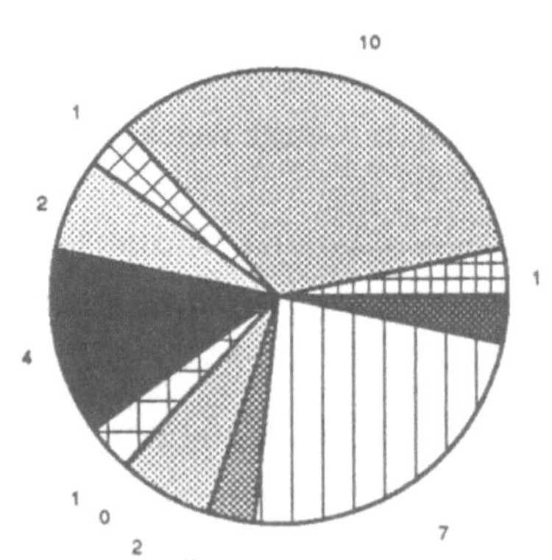

3. Projects where SME or Research Organization present

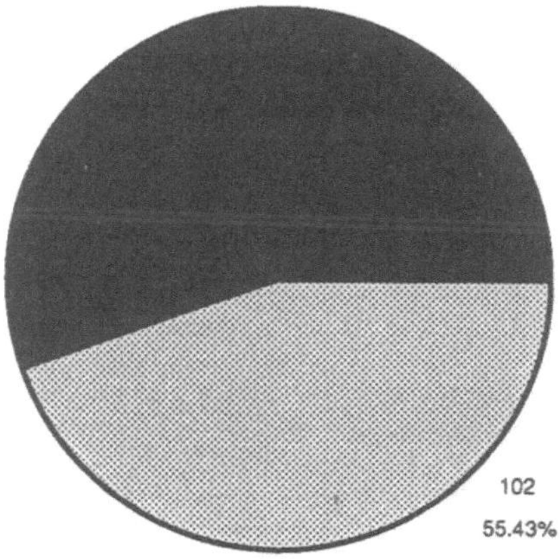

4. Projects containing SMEs

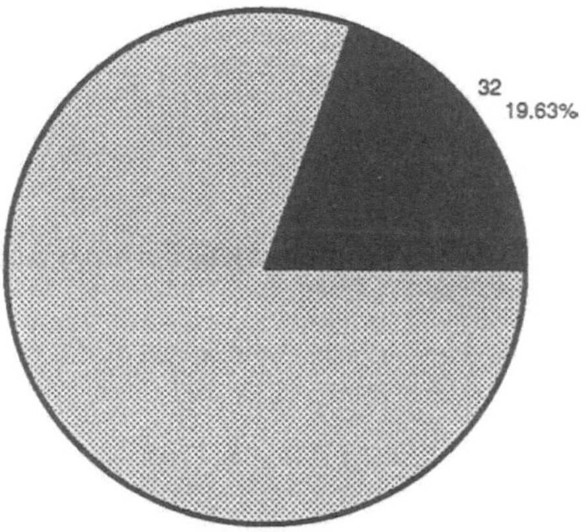

5. Projects where SMEs are prime proposers

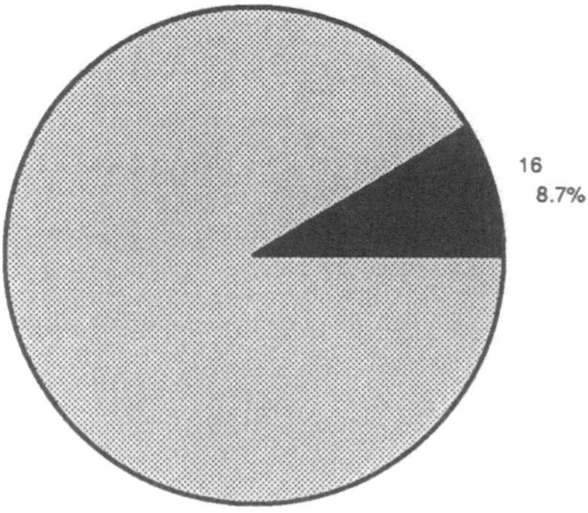

6. Projects where all industrial partners are SMEs

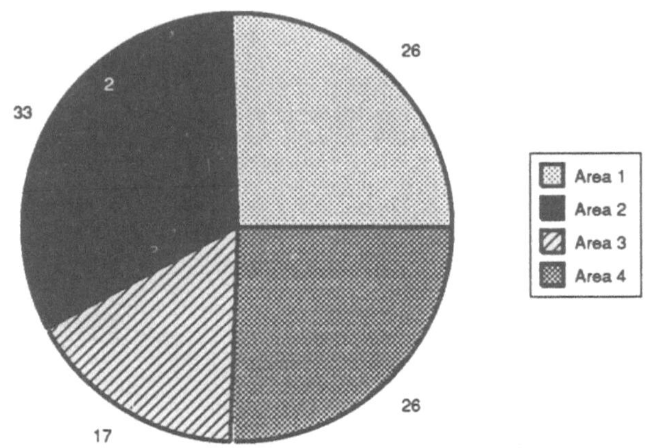

7. Projects containing SMEs broken down by topic area

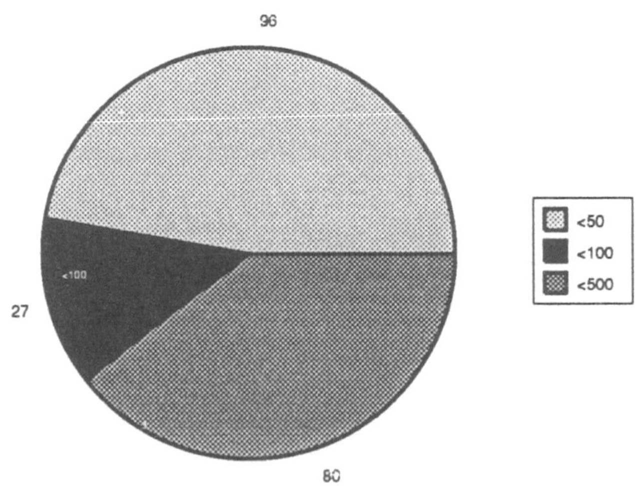

8. SMEs broken down by size

The SPRINT Programme – Innovation and Technology Transfer within and between Community Firms

B.B. Goodman

Introduction

SPRINT – the Strategic PRrogramme for INnnovation and Technology Transfer – is one of several Community programmes whose objectives include helping SMEs to reap the full benefits of the European single market with a view to 1992 and beyond.

SPRINT aims to overcome the differences of custom, language and law which characterize the European Community and to enable technology from one area or sector to be transferred to and used in others.

But what distinguishes SPRINT from most other Community programmes is that it focuses on encouraging innovation and technology transfer across Europe, rather than on other questions of real concern to SMEs such as R&D or purely commercial collaboration.

The objectives of SPRINT include:

- strengthening the innovative capacity of European producers of goods and services;
- promoting rapid penetration of new technologies throughout the Community;
- enhancing the effectiveness of existing regional, national or Community instruments and policies in the field of innovation.

Here it is appropriate to mention another Community programme with similar objectives, VALUE, which was approved in June 1989. However VALUE concentrates mainly on the dissemination and exploitation of the results of Community R&D programmes, which account for about 3% of all publicly funded R&D in the Community.

The scope of SPRINT is broader and its thrust is further downstream in the innovation process. Furthermore, while VALUE is part of the Community's R&D Framework Programme, SPRINT is not.

SPRINT was initially launched as a pilot operation in 1983. In that form it lasted until 1988 with a total budgetary commitment of nearly 20 million ECU. It was felt that this pilot operation should be directed at intermediaries or advisory agencies in the field of technology transfer rather than at direct support for some of the hundreds of thousands of SMEs themselves. In this way a much greater "multiplier effect" was thought possible.

Examples of the way in which SPRINT operates through intermediaries include:

- launching aid for the European Venture Capital Association (EVCA), which now has over 150 members and is fully self-supporting;
- launching aid for TII, the European Association for Transfer of Technologies, Innovation and Industrial Information;
- support for giving a European dimension to conferences related to innovation, so far attended by, altogether, over 10,000 delegates;
- the setting up of a European Community DESIGN prize;
- work on industrial property rights, including a comparative study of the cost of protecting them in Europe;
- support for less favoured regions, including a project for the introduction of robotics into Ireland and support for the development of the Greek and Portuguese patent offices;
- support for the EUROTECH project, involving collaboration between National Research Development Organisations from all Member States;
- support for the setting up of a European Association of Contract Research Organisations;
- and, especially as far as this theme is concerned, support for micro-networks for promoting technological agreements between firms in different Member States.

Sprint Micro-Networks

Human networks of advisory bodies, each with specialised knowledge of clients seeking or offering technology, and therefore capable of finding "matches" for the client firms of the network partners, are a most effective mechanism for technology transfer. Notwithstanding certain fears, no pressure is put on firms to harmonize their activities. On the contrary, firms with quite different complementary interests and resources are often very well placed to solve a common problem through collaboration.

The SPRINT Programme currently supports some 50 networks involving about 150 partners and about 600 firms, mostly SMEs. This rather cost effective result was achieved with an overall budget of just under 8 million ECU, or 40% of the total budget of the pilot phase of SPRINT. Following a recent call for proposals under the main SPRINT programme (see below), these numbers are expected to increase to about 70 networks with over 200 partners and substantially more firms.

Typical partners include:

- Technology Brokers
- Innovation Consultants
- Universities and Science Parks
- Regional Development Agencies
- Chambers of Commerce, etc.

The networks are each supported by the Commission through a contract which sets targets for the number of transnational technology transfer agreements between SMEs.

By the end of 1988, 185 such agreements had been signed and a further 121 are expected to be concluded this year (Chart 1).

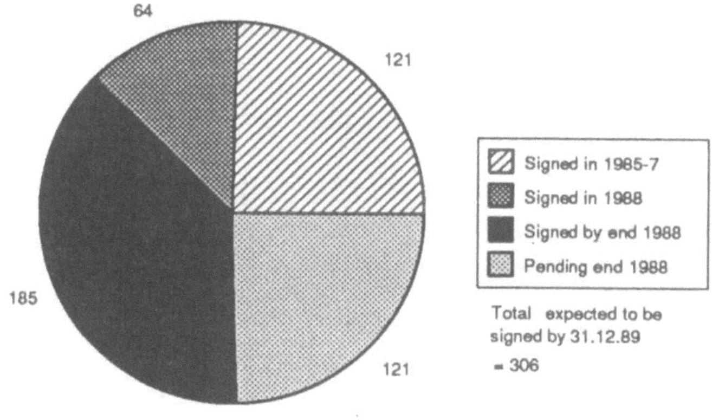

Chart 1: Evolution of the no. of transnational inter-firm agreements signed under SPRINT

Since each contract involves at least two firms, the total number of firms involved can safely be estimated to be at least 600. In the first 3 years the average number of agreements signed per year was 40 and we were encouraged to see this figure rising to 64 in 1988. As we moved along the learning curve we carried out a detailed survey last year into the type and nature of the agreements signed. Chart 2 indicates the types of agreement signed in 1988.

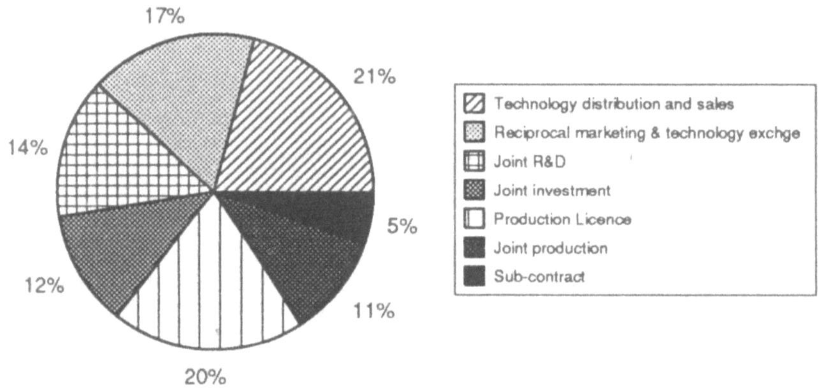

Chart 2: Breakdown of the types of agreement signed in 1988

The sectoral breakdown in Chart 3 indicates the wide range of sectors, from traditional to high-tech, which have been involved in the transnational agreements.

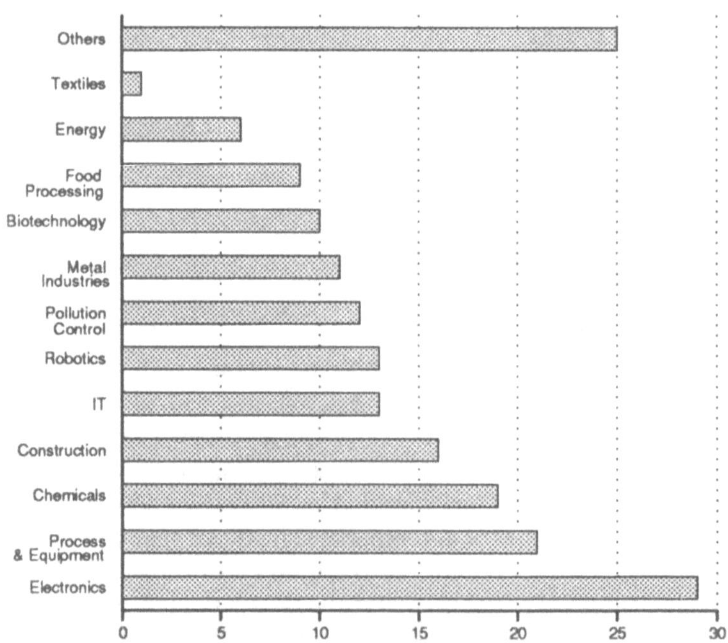

Chart 3: Sectoral breakdown of agreements signed by end '88

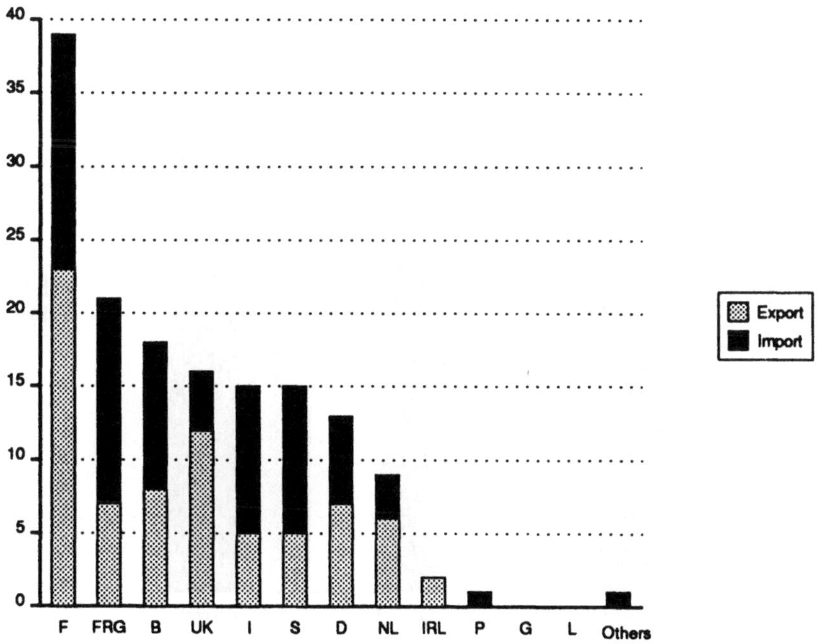

Chart 4: No. of technology exports and imports for the 64 agreements signed in 1988

Chart 4 shows the number of the transnational technology exports and imports, for the 64 agreements signed in 1988:

- France leads in the number of technologies exchanged;
- Italy, Spain and surprisingly, the Federal Republic of Germany, are indicated as net importers of technology;
- the UK and the Netherlands are net exporters;
- Ireland is a relatively small player but has potential in some sectors such as agro-food technology and biotechnology.

Chart 5 shows in a simple way that it takes micro-networks about two years to become productive. On two separate dates, at the end of 1987 and at the end of 1988, networks which had been in operation for at least two years accounted for less than half of the networks then in operation, and for less than half of the cost to the Commission of all networks, but they accounted for more than half of the agreements signed.

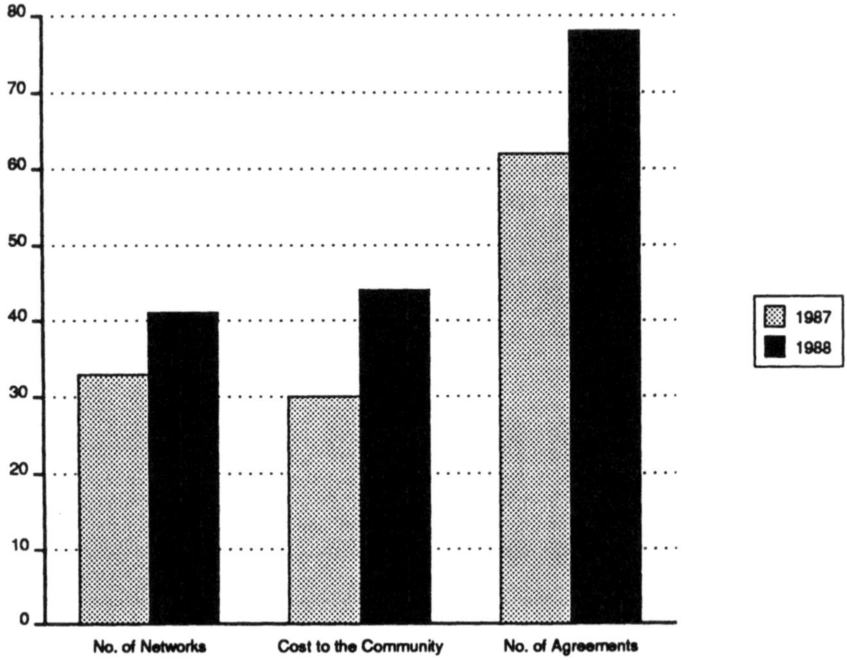

Chart 5: Data illustrating the fact that networks take 2 yrs approx to become productive

Three broad lessons have been learned in this work:

- The most important lesson is that there has to be very real commitment on the part of the partners to work together by setting goals and achieving them.
- The second important lesson concerns specialization, which may be geographical or sectoral. The networks which have specialised on a few client companies from a tightly focused sectoral grouping are achieving more results than those trying to do everything at once.
- The final lesson concerns the emergence of strong regional groupings such as those developing in Scotland, Denmark and Spain, and here the network partners – although they may strictly speaking be competitors – are prepared to meet and exchange opportunities which cannot be best exploited by the original recipient. This illustrates the importance of informal collaboration between firms which other speakers have stressed.

Joint promotional campaigns are also being discussed to increase awareness of SPRINT among SMEs in these regions. We want to encourage this building up of such regional strengths across the Community. Our experience shows that in order to achieve such a result there must be a focal point for the participants in each region.

I would now like to mention in particular a project, the intelligent greenhouse, which is described in greater detail in the booklet "Innovating across Europe".

The two companies, SEMIB-SOUDAGE and DANASA, were brought together via SPRINT partners in Bordeaux and Pamplona and a successful joint venture was developed.

In 1988 a third company in the field of information technology, TELEINFO of Ghent, was introduced by GOMOV, the Belgian partner in the network. The three way joint venture is currently planning a major demonstration greenhouse project near Bordeaux costing over 20 MECU.

The Main SPRINT Programme

In 1989 because of the results being achieved under the SPRINT pilot programme, the Council of Ministers approved the main SPRINT programme (1989-93) which was three priority lines of action:

- European infrastructure for innovation services:

  – establishing and strengthening intra-Community networks.
  This is an extension of the work already done under the SPRINT pilot programme.

- Specific projects for intra-Community technology transfer:

  This will involve providing support of various kinds for transnational pilot projects with industrial participation, involving the application of generic technologies to industrial sectors in less developed or declining regions in the Community. The emphasis will be on rendering the risks associated with the proposed technology transfer small enough, through appropriate studies and evaluations, for the firms concerned to commit their resources with confidence.

  The first call for proposals closed recently. Its aim is to select a few (10-12) large projects (up to 3 MECU) which will have a demonstration value so that the benefits of technology transfer in one case may be replicated in another area or sector.

- European Innovation Monitoring System:
  - better understanding of the processes of innovation;
  - increase cooperation between the Member States and the Commission.

  The central problem here is to arrive at a more exact understanding of the innovation process, so that the Member States and the Commission can keep track better of what is happening, notably under SPRINT, and take appropriate action in time.

  For many years it was considered adequate to measure levels of public expenditure on R&D. But that approach only measures one of the inputs to the innovation process, whereas it is no less necessary to measure the outputs. In the United States there is the National Science Foundation's publication SCIENCE INDICATORS, and there are initiatives in the OECD, but so far little exists at Community level.

The budget deemed necessary for the main SPRINT programme is 90 million ECU.

I have tried to give an outline of the contribution of the SPRINT Programme to the cooperation phenomenon but it is only possible to touch on some of the main points of the Programme which is complex because it addresses a complex problem. More information can be found both in the leaflets "Innovating across Europe" and "Introducing Innovation" and in the DG XIII/C quarterly Newsletter "Innovation and Technology Transfer".

The Programme for Cooperation – Policies and Programmes for European Partnerships

G. Heikens

Introduction

The fact that this book includes contributions from a large number of representatives ofthe European Commission, clearly underlines the interest of the Commission in the phenomenon of cooperation. This interest has been growing simultaneously with the progress made on completing the Internal Market. However it is older than the "White Paper for the Completion of the Internal Market". Before the Commission published this White Paper, a number of Community initiatives already promoted cooperation, such as the activities of the Business Cooperation Centre.

In this paper I will discuss:

- why should companies cooperate?
- how to develop a partnership and;
- what are the methods of DG XXIII (Enterprise Policy) for stimulating partnerships.

Why Should Companies Cooperate?

By putting this question one should clearly keep in mind that cooperation in itself can never be the objective. It is only a means, a strategy to attain other objectives such as: increase of the market share; getting new technologies at your disposal and; strengthening the financial basis of your company. So having defined the company objectives, you can choose the appropriate strategy. Cooperation can be such a strategy because it can improve the competitiveness of European industry, which is important in European as well as world markets, in particular the United States, Japan and the Newly Industrializing Countries (NICs). By cooperation you get access to new markets, new clients, new suppliers, new techniques, new technologies, new financial means etc. By new, I mean new to you. Usually they exist already somewhere. By cooperating you use existing methods, making it a strategy that is cost and time efficient.

I would also like to make clear, that when I am talking about cooperation, I am not only referring to joint ventures, mergers and acquisitions.

On the contrary, in particular for small and medium-sized enterprises, less complicated forms may be more suitable. For example, mutual distribution agreements, franchising, common marketing, export-clubs, licensing, co-manufacturing, sub-contracting, transfer of technology, common research and development, etc. These are all forms of cooperation which are useful for SMEs. Our initiatives emphasize these forms of cooperation in particular.

How to Develop a Partnership?

One can distinguish the following phases in developing cooperation:

a. Preparation. During this phase you analyze your company, its strong and the weak points, your competitors, your suppliers, markets, etc. with the objective of fixing your targets and defining the strategy. If the conclusions of this analysis point to a cooperation approach, the following phase commences.

b. Definition of the form of cooperation and a profile of the desired partner.

c. Partner-search. Through various channels you or a consultant, acting on your behalf, identify some interesting and interested partners.

d. Negotiation. You start negotiations with the selected company. During this phase you define the common target, the juridical form (e.g. European Economic Interest Group), the tasks of each partner, responsibilities etc.

e. Execution. This might be the most difficult part. Now the intentions and plans have to be transformed into concrete actions. Depending on the type of cooperation envisaged a number of problems may arise.

Having described, in a nutshell, the process of setting up partnerships, it will be clear that it is a complicated, time-consuming process, which requires full commitment by the management. Do not treat it as a side-line activity, which you will look after if you have a quiet moment – firstly because you don't have quiet moments and secondly because ;the success of cooperation is related to the driving force of the person behind it.

Initiatives of DG XXIII to Stimulate Cooperation

Cooperation is a matter of business to business relationships. Therefore the role of the Commission must of its nature be limited. It can provide services to the business to facilitate some of the phases, in particular if it comes to transnational cooperation. On the other hand it must be clear that it can never take the place of an entrepreneur.

The fields where the Commission can play a role are

- the legal framework; e.g. juridical forms and competition;
- the phase of partner-search and;
- training and information.

I will confine my presentation to the initiatives developed by DG XXIII in the latter two fields.

We developed the following:

a. BC-NET Business Cooperation Network.

BC-Net has two main features.
Firstly it is a network of about 400 business advisers. Secondly it is an automated system to match supply and demand for cooperation. It is a mechanism through which the business-person can identify interesting contacts in the other Member States. About 5000 cooperation profiles are stored in the computer. BC-NET receives about 50 profiles per day.
We are, with the BC NET members, constantly developing the system and the services it is providing. Linkages with countries outside the Community will be realized.
We are looking for possibilities to make extra use of the network and the facilities. We are particularly interested in "market-demanded features". One option is to link the BC-NET members with various databases in Europe.

b. Subcontracting.

The Council of Ministers adopted in September 1989 a resolution to stimulate transnational subcontracting. The general lines of this policy are:

- to create an environment suited to subcontracting;
- to improve the flow of information and communication between main contractors and subcontractors;
- to promote subcontracting and partnership relations.

Among the actions in this field are the publication of a practical guide to legal aspects of industrial subcontracting; publication of terminologies and follow up work on the creation of a European Information Centre on Subcontracting.

c. Cooperation projects.

The best known action in this category is Europartenariat, which took place in Ireland in 1988. It is a programme of DG XXIII, Enterprise Policy, and DG XVI, Regional Policies, to stimulate partnerships among enterprises of a less favoured region of the Community and the other Member States. The project has developed rather quickly. In 1988 we brought together 120 Irish firms with 100 industrial companies and 80 business organizations. In 1989 more than 200 Andalusian companies got in contact with more than 450 firms plus 150 business representatives. Wales will be the targeted region for 1990. The two contact days will be held in Cardiff on 14 and 15 June. A medium term programme has been developed. Europartenariat 1991 will happen in Porto, Portugal while Greece will host the event in 1992. We are examining the possibilities of using the formula of Europartenariat to select companies in one region, collect cooperation proposals, partner-search in another region and bring the businesses together, in order to stimulate cooperation among firms from two or a limited number of regions.

DG XXIII is prepared to discuss with interested organizations the possible support from the Commission for setting up these types of programmes. We are looking into the possibilities of applying a similar formula to actions with countries from outside the Community.

d. Training

To prepare SMEs and in particular SMEs who are new-comers to international cooperation, we are taking two initiatives to help them to prepare themselves for the process of setting up partnerships.

The first one will be the publication of a guide on cooperation. It explains the various phases, which I mentioned earlier and lists the items an entrepreneur should be aware of when he wants to set up a cooperation arrangement. A second initiative is an action called "Prepare the management of SMEs in Europe for 1992-Strategic Management Training". In Appendix 1 to this paper you can find the list of organizations in Ireland who are in charge of the execution of this action. The Commission is studying the possibility of setting up a course in "Cooperation".

Conclusion

In this paper I have illustrated the initiatives of DG XXIII with regard to a strategy of "Cooperation". Our future activities will be concentrated upon the further strengthening and extension of our existing approaches in the indicated direction. It is our wish to develop these approaches in close collaboration and consultation with users in order to ensure that we are on the right track.

Appendix 1

Training Scheme: To Prepare SMEs for Europe 1992

List of Organizations in Ireland in Charge of the Action:

Irish Management Institute,
Sandyford Road,
Dublin 16.

Marketing Centre for Small Business,
University of Limerick,
Plassey Technological Park,
Limerick.

Byrne Lowe and Associates,
21 Northumberland Road,
Dublin 4.

Irish Productivity Centre (IPC),
IPC House,
35-39 Shelbourne Road,
Ballsbridge,
Dublin 16.

Commission of the European Communities,
DG XXIII, Att. Miss J. Gaudin,
Rue de la Loi 200, ARLN 5/70,
1049 Brussels.

The Structural Funds – The Means For International And Interfirm Linkages

J. Mulcahy

Within the rather wide brief I have been given I have selected five issues to address, as follows:

- the realities of partnerships
- European Community Business and Innovation Centres
- Europartenariat
- European Economic Interest Grouping
- reform of the Structural Funds

The choice of title for this book, i.e., the Cooperation Phenomenon, is interesting and thought-provoking.

"Cooperation" in practice is interpreted in different ways by the people involved, with a certain tendency to regard it as a one-way process.

"Phenomenon" implies a sense of the remarkable or the unusual.

At the risk of stating the obvious, it is vital that the partners in any cooperation venture have a clear mutual understanding of their fundamental aims and motivations in entering into such a unique relationship – how they are related and how they are different. The operational phase of such ventures often highlights a lack of understanding with regrettable consequences and recriminations.

Public/private partnerships are frequently quoted as the way forward in local and regional economic development, but it is important, to understand that whilst the primary object of the public sector is the attainment of the public good, both socially and materially, the primary object of the private sector is ensuring commercial success and profitability. To the extent that both sectors can achieve their primary aims, local and regional economic development will progress.

In wider partnerships, the diversity in the makeup of prospective members, their varying interests and the difficultiessu of ensuring sucessful combination may pose considerable problems. The variety is evident in the issues addressed in this book, viz. small firm to small firm cooperation, large firm to small, established company vis-a-vis young hi-tech firm, smaller nations, smaller economics and so on.

When an international overlay is introduced with legal, financial and language difficulties, the complexities become daunting, especially at the lower end of the scale and in the less prosperous regions.

In this connection it is interesting to note the findings of a study carried out for DG XVI on international cooperation between SME's and the impacts on indigenous regional development in Ireland and Greece. The team consisted of Professor Lucey of University College Cork, and Professor Demathas, the Pantheios, University of Athens, coordinated by Professor Hannequart of the Universities of Mons and Louvain. They point out that companies' objectives in transnational agreements are rarely realized or only to a limited extent through what are less acceptable arrangements. The costs of negotiation and implemention are a major inhibiting factor.

Business and Innovation Centres' experience with the implementation of the Community's Structural Funds, especially the European Regional Development Fund (ERDF), emphasizes that, self-sustaining and resilient local economies must be based on indigenous leading sectors consisting of knowledge-based firms that can compete and export internationally. It is also clear that while there may be sector inputs from outside the region, they will be in support of and not in place of indigenous potential.

There may be an array of local and national assistance available in this context but the evidence and experience of the last fifteen years of Community regional development policy reinforces the mounting anxiety that present approaches are not sufficient, and should be more dynamic and responsive to business and market pressures and indeed to local demands.

Efforts to counter the severe redundancies which followed in the aftermath of the restructuring of the coal and steel industries in the early 1980s and the need for a system to generate an ongoing flow of people and projects on which financial assistance could be brought to bear eventually led the Commission to the development of the "business and innovative centre" concept.

Business and Innovation Centres (BICs), in the promotion of which the Commission acts only as a catalyst, are public/private cooperative ventures designed to respond to the partnership and service requirements already outlined. The "total systems" approach which these Centres represent is based on partnerships between public business development agencies and private sector interests cooperating and combining to use and finance the Centres in complementing and widening the scope to their activities. The functions of the BICs include identifying and selecting entrepreneurs and technically-based business opportunities (indigenous potential) and translating them into economy-leading businesses by providing, in an easily accessible way and location, the comprehensive range of skills development, business planning and access to finance services. These are essential if upmarket business people and possibilities, of which Ireland has an unparalleled endowment, are to be brought into the local SME sector. Start-ups and existing SMEs require similar assistance to succeed, in particular in relation to cooperation with client firms associated with other Business and Innovation Centres throughout the Community and beyond via the Centres' international network.

It is the intention of the Regional Policy Directorate General to use the Business and Innovation Centres and their network as the vehicle for pursuing further innovative SME development pilot actions, such as the present Seed Capital Scheme.

Commission departments have been keenly aware of the advantages and problems of SME cooperation and have helped, and continue to help in a variety of ways.

Although it is outlined in much more detail by Mr Geerth Heikens of DGXXIII, I should, nevertheless, like to refer briefly to the SME partnership project entitled "Europartenariet" or European Partnership which is essentially aimed at generating cooperation agreements between companies selected in less prosperous Community areas and seeking to find and bring them together with prospective partners from the more prosperous member states.

The Regional Policy Directorate General XVI collaborates closely with the Directorate General responsible for small and medium-sized business on a community-wide basis (XXIII) to fund and operate this project. The first event was held in Dublin in 1988, the second in Andalusia in '89 and the third in Cardiff on June 14-15, 1990.

Experience with this exercise underlines three fundamentals requirements which any SME interested in cooperating internationally or otherwise must possess:

- an innovative technology, product or service which can form the basis of a commercial cooperation agreement, joint venture, licensing arrangement, financial investment etc.
- a company management structure on both sides capable of implementing an agreement.
- a strong willingness by both managements to undertake the tasks and succeed.

As previously mentioned, the incompatibility of national legal and financial requirements is a severe deterrent to international SME cooperation. Very considerable Commission attention and effort is being devoted to the elimination of these barriers particularly by DGXXIII and XV (Financial Institutions and Company Law). The European Economic Interest Grouping (EEIG) Regulation No. 2137/85 which came into force on 1 July 1989 allows companies within the Community to engage in joint activities such as research and development, purchasing, production and selling, specialized services, computerized data processing and the formation of multidisciplinary consortia in the construction industry to tender for public or private contracts. The framework and rules laid down in the Regulations are particularly designed to cope with the needs of small and medium-sized companies.

The formalities involved in forming an EEIG are simple: a contract is concluded between the members and submitted to the official Member State registry. Registration confers full legal capacity on the EEIG throughout the Community and even outside. The grouping enjoys neutrality in respect of its profits, both for commercial and tax purposes, in so far as its profits or losses are taxable only in the hands of its members.

I am sure you are all now familiar with the reform of the Structural Funds and the efforts being made to implement the provisions.

The primary object is to achieve better economic and social cohesion in preparation for 1992 by reducing existing regional disparities. The Funds, are designed to encourage local development initiatives and support cooperation in a flexible and responsive way. Specific provision is made for a global grant mechanism for regional intermediaries for in local development initiatives, and for pilot schemes to pool experience and encourage development cooperation between different Community regions.

My work programme during the coming year includes concentrating attention on networking to stimulate, mobilize and coordinate the efforts of bodies and agencies, both public and private, active in local economic development to ensure more effective mutual benefits for all concerned.

May I conclude by stating that cooperative action is clearly identified by the Community as a key factor in the development of the less prosperous regions, but achieving it requires recognition of the fundamental change required and the discarding of outmoded systems and attitudes.

The EUREKA Initiative for West European High Technology Cooperation

O. Meyer

The EUREKA Initiative

Launched in 1985, EUREKA has changed the face of R&D cooperation within Europe. It is an innovative tool helping Europe to master and exploit the technologies which will prove decisive in the race for competitiveness.

EUREKA's 20 members are the EC and the EFTA countries, Turkey and the Commission of the EC.

Its aim is to stimulate cross-border cooperation in order to heighten Europe's productivity and competitiveness on the world market.

Its action sweep covers all scientific and technological fields, which means that no single technological field is prescribed as a limiting area. However, since there is no international financing within the Eureka Initiative, all financing must be found on a national basis, (private or public as the case may be), and the project therefore might be requested to belong to nationally-defined priority areas.

"Bottom-up" is Eureka's ground rule. Participants have full responsibility for defining and implementing their scientific and technological cooperation projects. They are their own judges of the best course towards new markets for Europe.

In order to become a EUREKA project, the venture in question must:

- involve at least two partners from different EUREKA countries
- use leading-edge technology
- aim at securing a significant technological advance in the product, process or service concerned.

From an administration point of view, the Initiative is supported by a political body, the 20 national project offices and an international secretariat.

The EUREKA offer to the project participant

1. Networking Resources

- You have a project proposal, but no partners?

- You are already involved in a project, but it demands specific skills, necessitating a further partner search?
- You do not have a specific proposal, but would like to enhance your technological know-how within a European framework?

  The EUREKA National Project Coordinator for your country can put you in touch with the right people throughout Europe. You can consult the EUREKA database, which lists more than 300 co-operation projects with almost 1600 participants and will also be made a tool for active partner search.

2. A Meeting of Minds: Supportive Measures

- Your project is in need of new joint industrial standards?
- It faces technical obstacles to trade?
- It is hampered by barriers to public procurement?

  EUREKA offers a forum for direct dialogue between participants and Governments or European authorities. You can put your case and state your requirements through EUREKA to those who have the authority to act.

3. Access to Public and Private Funding

   Participants are expected to raise the funding required for the implementation of their projects. However, Governments of EUREKA member countries can decide to provide financial backing for projects. In such an event, they set the level of funding and eligibility conditions. More often than not, you will find priority for EUREKA projects in public funding schemes.
   EUREKA can also act as a contact point for participants seeking access to private funding.

4. The Hallmark of Professionalism

   If you project receives the EUREKA seal of approval, it can wear the EUREKA label – an internationally recognized hallmark of professionalism.

5. A School for Cooperation

   Even if crossborder cooperation sounds interesting enough to enhance the speed and efficiency of technological progress, cooperation as such is an art one has to learn. The EUREKA network has observed hundreds of cooperation projects and is able to assist you with advice and some checklists when it comes to organizing your cooperation plan.

The EUREKA Project Portfolio so far

The EUREKA Initiative is just four years old, which constitutes very few years in technical research and development activities. The project portfolio has consequently not reached any stable level yet (if, indeed, it ever will do so) but is developing steadily.

The projects which, during a year, reach EUREKA status are announced at the yearly Ministerial Conference, which is usually arranged in the month of June.

All announced projects up to and including the latest Ministerial Conference in Vienna, total 297 with a budget of 6.4 b ECU and almost 1,600 participants are represented in Appendix 1 to this paper, as follows:

Chart 1 represents the participation per country details
Charts 2, 3 & 4 represent the distribution by area
Chart 5 represents the cost distribution of the projects
Chart 6 represents a classification of the participants

The Irish Participation

Ireland – as can already be seen from Chart 1 – participates, as yet, in "only" 9 projects, which critics might consider a rather small number. It might, however, be advisable to take a closer look at the question of how much this "little" volume can actually achieve and represent in value.

Charts 2a and 3a represent the Irish participation per area, where lines "COM"munication and "ENV"ironment will be worth studying. In both these cases, Irish participants are found in "only" 2 out of 19 and 32 projects respectively, but participate in precisely those projects, which constitute one third of the total value per technological area. The lesson to be learnt: even small participants can belong to an interestingly large part of a technological area.

Chart 1a presents the number of joint project participants. Ireland meets in its "mere" 9 projects. Again, the multiplying effect is quite astounding, even though easily explainable, when one considers the simple fact that one single project can have many participants.

Possibilities for the Smaller Economy

As touched upon already in the previous section, the first and dominant argument for smaller economies to seek international cooperation, should be the multiplying effect which can be achieved. One can actually reach many partners with relatively few projects, one can actually act as a financially small partner in large projects; as the Charts 1a, 2a and 3a demonstrate.

Of course, this policy should not be pursued in an unreasonable manner. The reputation of the smaller economy must be kept high in the sense that quality is always demonstrated. The larger economies will always appreciate that quantity is not at hand, but lack of quality will hardly be accepted in the long run.

The variety and number of contacts that cross-border cooperation can and will generate is another benefit of special value to the smaller economy. The smaller the economy is at present, the greater the relative value of another contact, another learning opportunity, another market possibility. Again, however, the quality aspect should *never* be disregarded.

To start an international cooperation venture must at any rate be regarded as a cost-consuming exercise. To do this, with only "foggy" reasons, with inadequate resources as to competence, personnel and finance and with the initial drawback of being a smaller economy, cannot be called anything but disastrous – in a short and longterm perspective.

Experience has shown that the EUREKA label on a cooperation venture will be regarded as a token of legitimacy implying quality, honesty, commitment, proper planning and adequate resources on the part of the participating units. This token will not be supplied if the prerequisites are not demonstrated.

In conclusion, there are obvious possibilities for smaller economies to utilize international cooperation: carefully planned, it may result in remarkable multiplying effects, but quality has to be maintained. This calls for identifying the national centres of excellence, which every country has available, and encouraging them to participate and thereby to stand out as the builder of the country's reputation.

Possibilities for the Smaller Participant

The smaller participant is well advised to consider almost the same arguments as formulated above for the smaller economy. The arguments come, however, in a different order.

The single participant – it might be a research institution or a company – must first of all evaluate his own financial future. Companies must always, and with no excuse, consider their profit situation (because that is the base for their existence) and so called non-profit organisations might have an even more difficult task in avoiding profit and simultaneously avoiding losses.

The smaller participant will always appreciate the legitimacy that a EUREKA project possibility can provide him with in his contacts and partner-seeking process, especially when it comes to contacting (very much) larger partners. As a personal remark, the author has to admit his amazement at the degree of legitimacy the EUREKA label can give to the small participant. Nothing appears to be left of the "not invented here", "small inventor – big company" or similar all too well-known syndromes.

The next step for the smaller participant is to very carefully analyze the usefulness of the cooperative venture. One plus one must result in more than two in order for two partners to feel content with the trouble of cooperation. This is valid in general. The smaller participant must, however, secure the even more difficult task of getting 1 + 20 to become greater than 21 (where he would represent the 1 and his big partner the 20) and also, that his bigger partner understands and appreciates the advantage for himself.

Cooperation means sharing. Sharing of duties and rights, sharing of costs and income, sharing of the positive and the negative aspects of any joint endeavour. By nature, the larger partner will be able to prove more easily that he can carry costs, that he has resources.

The smaller participant must obviously work harder and more carefully to prove his point and his ability. But if he does and is successful, he will build a reputation of great value, giving him improved possibilities the next time.

In conclusion, the smaller participant can receive legitimacy through EUREKA for a first international cooperation venture, but will have to build up his reputation by quality actions so that the legitimacy of his second venture can rest on the previous success.

The Norwegian Participation

Addressing a predominantly Irish audience – actually the first time in the author's life – one may be allowed to round off with a little push from constructive competition. Is it not the case that the mentality of the Irish gives them unusual strength to take up a competitive challenge, to withstand difficulties and to "make it" at last?

With this conception as a backcloth, the equivalent figures as represented above for the Irish EUREKA participation (see Charts 1a, 2a and 3a) are presented for the Norwegian participation.

Charts 2b and 3b therefore represents the Norwegian participation per area, whilst Chart 1b represents the number of joint project participants that Norway meets in its 34 projects.

Norway's population is only a little bit larger than Ireland's. Otherwise there are, of course, lots of differences, which can explain almost anything.

But Ireland's 9 against Norway's 34 EUREKA projects: shouldn't something be done about it?

Appendix 1

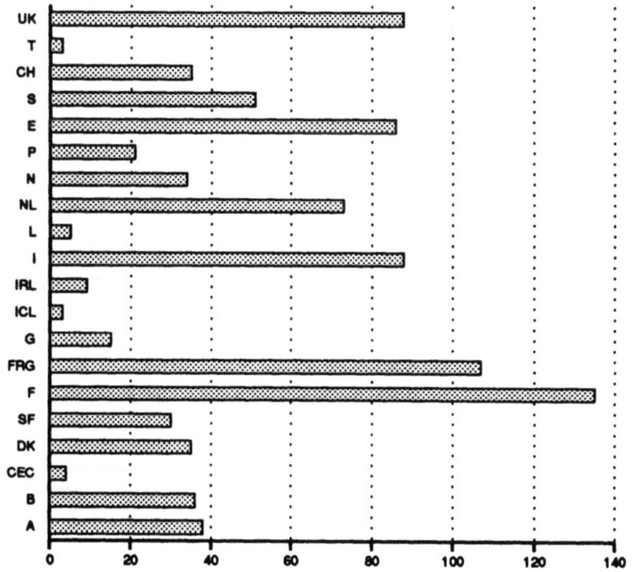

Chart 1: Project Participation by EUREKA members

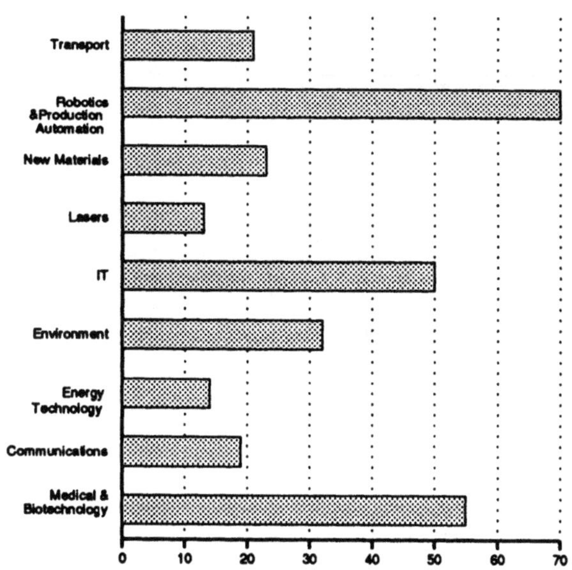

Chart 2: No. of Projects by area

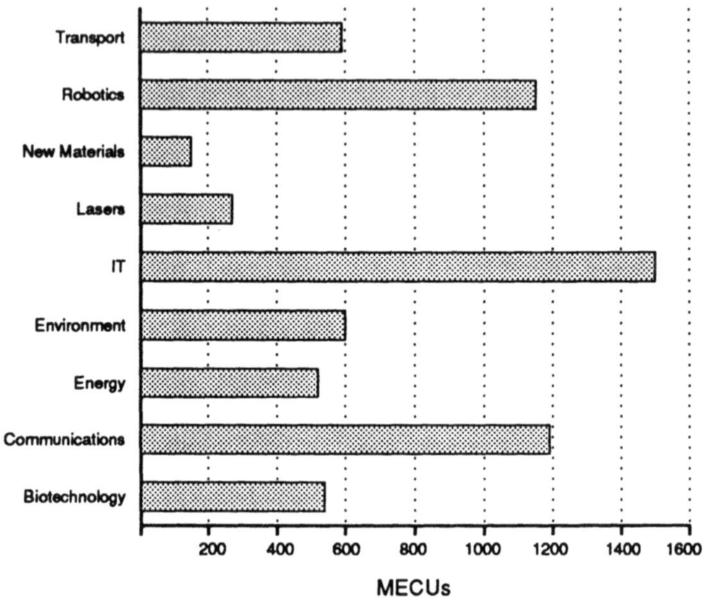

Chart 3: Total estimated cost by project area

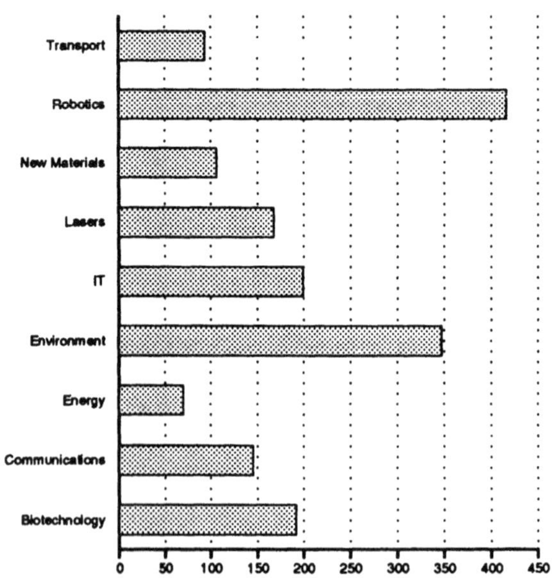

Chart 4:Total no. of participating organizations by area

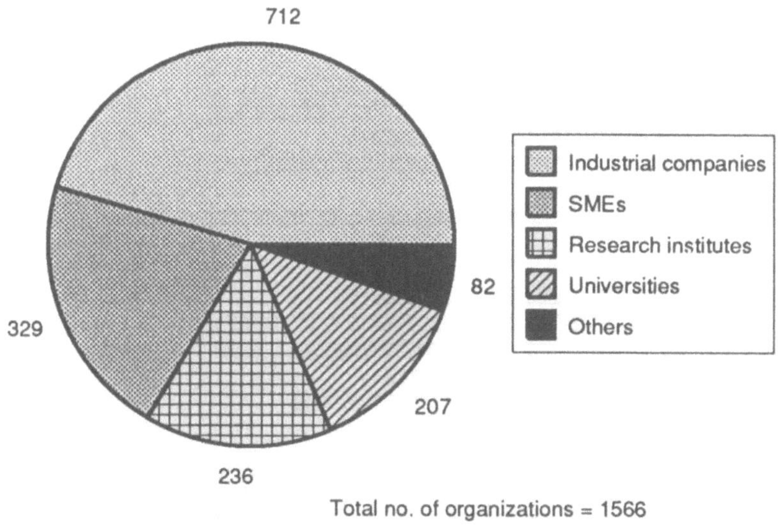

Chart 5: Participation organizations

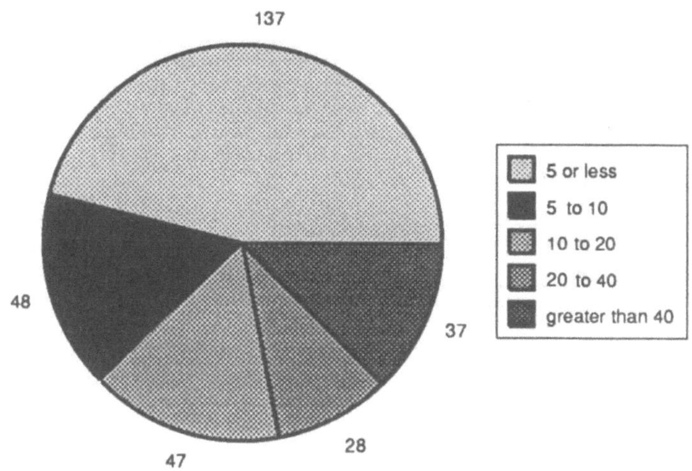

Chart 6: Financial size of the projects

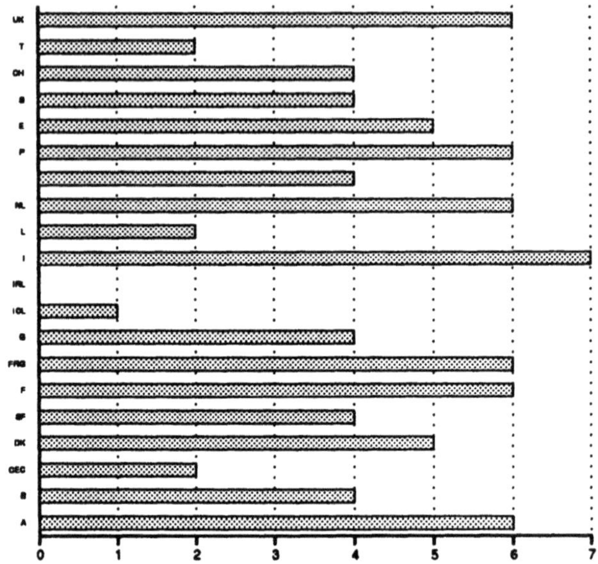

Chart 1a: Joint project participation between Ireland and other EUREKA members vs. total project participation

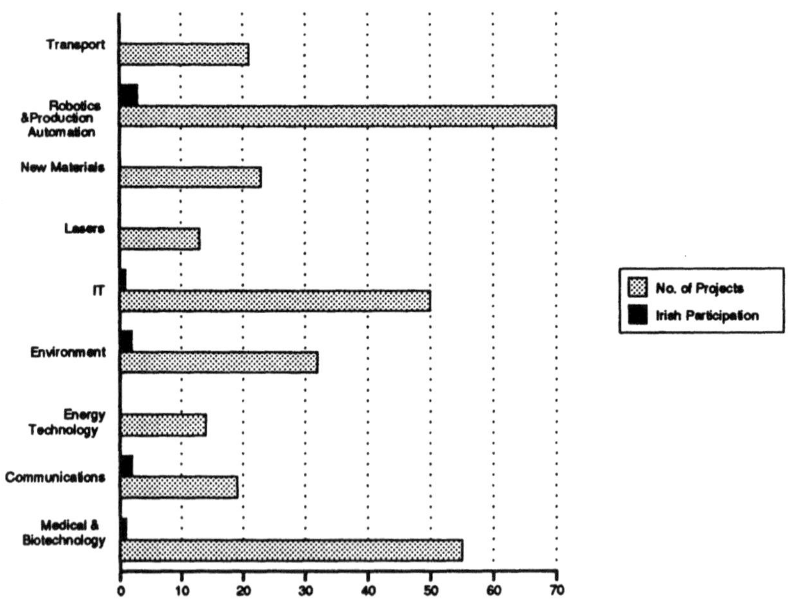

Chart 2a: No. of projects with Irish participation vs. no. of all projects

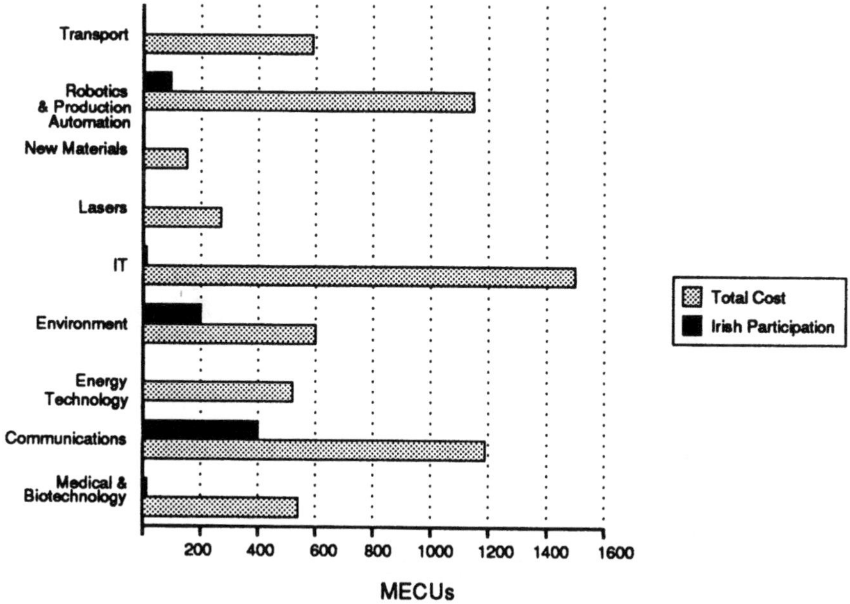

Chart 3a: Total cost of projects with Irish participation vs. total cost of all projects

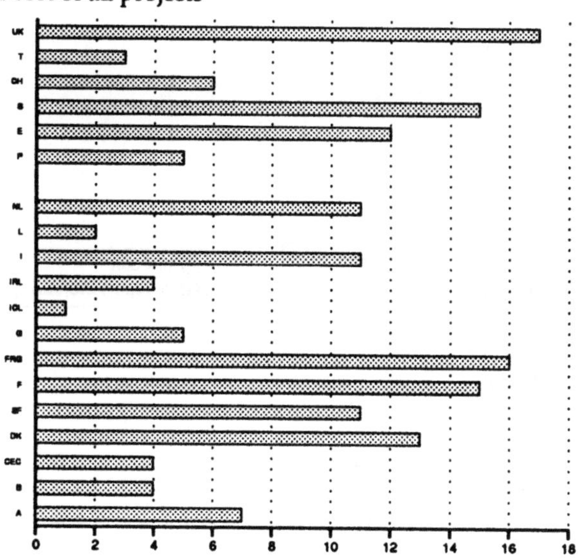

Chart 1b: Joint project participation between Norway and other EUREKA members vs. total project participation

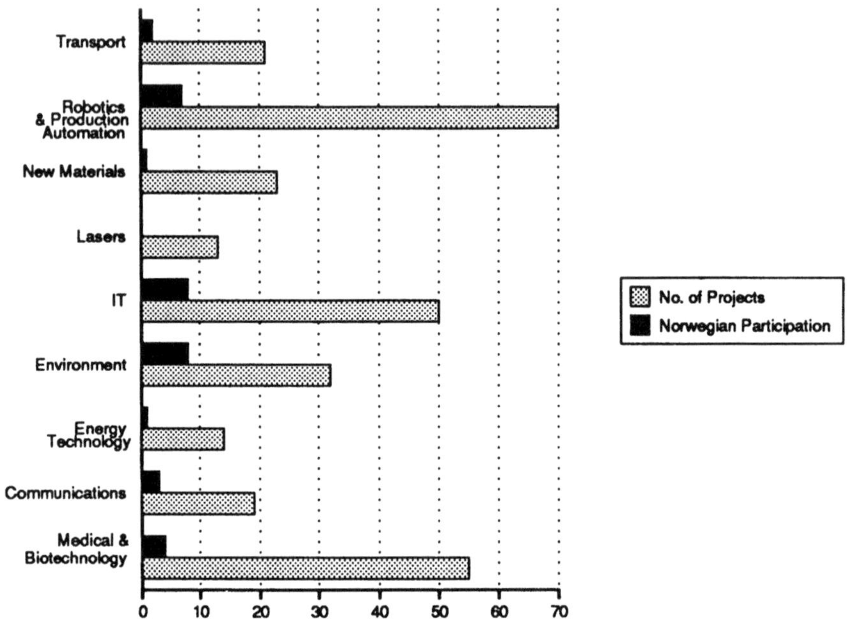

Chart 2b: No of projects with Norwegian participation vs. no. of all projects

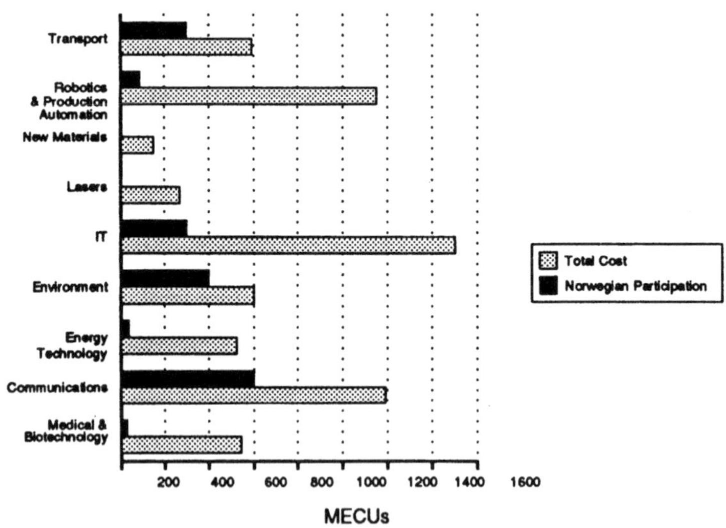

Chart 3b: Total cost of projects with Norwegian participation vs. total cost of all projects

Section 3: Small Countries and Economies – What is Different?

Small countries and economies, including both small national economies and the regional economies of large countries, are not just smaller versions of the larger economies. On the one hand, they are culturally and socially distinct and have some advantages of flexibility and responsiveness. On the other, they are structurally weak in terms of having the scale of technological, market, financial and managerial resources and capabilities now necessary for competitiveness in increasingly globalized markets. These macro level characteristics and disadvantages are both reinforced by and reflected in the preponderance of small firms in the structural makeup of most of the smaller economies.

This has an impact on the nature of the cooperation activities in which the firms of these economies tend to become involved. Because of the restricted and highly competitive nature of the domestic market, horizontal relationships, i.e. between competitors, are obviously almost non-existent. Where horizontal relationships do exist they will be with foreign-based partners to access distant markets or vice versa. The scope for vertical cooperation i.e. with suppliers and customers may also be somewhat restricted, particularly outside the areas in which the small economy has tended to specialise. In addition to the lack of studies referred to in a number of the papers in this section, small economies do not lend themselves, within their own territories at least, to a high level of formal cooperative activity. An external orientation can obviously assist in encouraging the cooperation process and this is why involvement in European Community Programmes can be of disproportionate importance for small firms and small economies. It is their structural context rather than anything inherent in the nature or makeup of the small firm in the small economy which militates against their involvement in the cooperation phenomenon, at the formal level at least, and this needs to be addressed by broad policies and programmes if the level of involvement is to be increased.

There is some evidence that informal networks as a means of acquiring knowledge are as well or even better established in the small firm/small economy environment as they are in that of their larger counterparts. This may partly account for the rather low level of formal licensing found in Portugal and Ireland for example (see Section 1). While some of the factors influencing the propensity to share information in this way are obviously culturally-based, the Danish authorities (see the first paper in this section) obviously believe that the level of such interaction can be increased through positive action on the part of the national and perhaps international authorities.

The first paper in this section raises the issue of "organized" markets, as opposed to the classical "free" market concept, and their importance for informal cooperation in a small country context. The authors analyse the nature and significance of user-producer relationships in the innovation and cooperation process and how they lead on to horizontal relationships between competitors, even in small domestic markets. The policy response of the Danish Government in adopting new technological and industrial policies, aimed directly at encouraging and enhancing cooperation and particularly networking, makes interesting reading for policy-makers elsewhere.

The second paper in this section draws on the experience of the author in evaluating the impacts of the EC ESPRIT Programme and particularly its impacts on small firms and small countries. It shows that "small entities" can bring a unique contribution to international collaborative projects but that conscious decisions must be taken at firm, national and international level to ensure that small firms and small economies get an equitable share of the results.

In the final paper the editor takes an overview of the cooperation phenomenon and its significance, mainly in the Irish context. The focus on R and D rather than innovation both in the literature and in national and international policies and programmes is questioned, and the need to take account of all the actors involved is underlined. Finally some emerging evidence is put forward that "ownership" issues are not now necessarily of primary importance in intra and inter firm cooperation, nationally and internationally, but a capacity to capture and appropriate the benefits, both direct and indirect.

Interfirm Cooperation in Small National Systems of Innovation

B. Dalum and B. -Å. Lundvall

Introduction

The themes of this bookincludes fundamental policy questions. Is the present institutional set up of small economies a good one? Should the present mix of market, organization and cooperation be changed in order to enhance the performance of small firms and small economies?[1] In order to answer these questions it is important to get a realistic understanding of the present institutional set up. Which role does interfirm cooperation play in small countries and to what extent are small firms involved in this kind of cooperation?

Empirical research on interfirm cooperation agreements appears to have been biased towards large firms, and firms from large and medium sized countries. Much of the analytical effort has been focused upon cooperation between multinational firms using cooperation as an instrument in a global strategic competitive game (for a survey from the point of view of "international business", see Contractor & Lorange (eds.) 1988). Does this imply that small firms do not take part in cooperation? Is cooperation relevant only to large firms and countries? In this paper we shall analyze the institutional set up of the economy as a framework for innovation. In this perspective cooperation turns out to be a more fundamental phenomenon than assumed in the literature on interfirm cooperation. Registered cooperation agreements appear to be only the tip of the iceberg and may, thus, not necessarily be the best indicator of actual cooperation.

However, we still assume that the rapid growth in the number of registered cooperation agreements reflects the increasing importance of cooperative behaviour in general. In order to assess different policy alternatives it is important to know what lies behind the current growing importance of the cooperation phenomenon. We shall propose that it

1 Concern about the competition/cooperation mix is a current theme in large countries also. In a recent report a group of distinguished US experts (the MIT Commission on Industrial Productivity) concluded that US competiveness should be strengthened through a stronger element of cooperation (Dextouzos et al., 1989)

represents a transitory & permanent change, reflecting that we are now entering a new era where interfirm cooperation will become a more adequate framework for innovation.

Finally, we discuss the policy implications for small countries and, especially, small firms in small countries. It is argued that policy makers at the national and European level must take into account differences in the economic structure as well as differences in the institutional set up between countries. Of special importance is the presence in the national economy of large domestic or foreign multinational firms.

User-Producer Interaction

One way to illustrate the general importance of cooperation for the process of innovation is to focus upon product innovation and user-producer interaction. Our point of departure is that a large number of innovative activities address the needs of users not formally involved in the process. This creates a mutual problem of information on the needs of potential users and the user must get access to information on the use value characteristics of the new product.

This mutual information problem cannot, possibly, be solved in a neoclassical world of pure markets where all agents are assumed to be anonymous to each other and where the only exchange of information takes the form of price signals. The alternative transaction cost approach, originally launched by Williamson (1975), would typically point to vertical integration as the rational and probable outcome in such cases. None of the two approaches can explain why private firms use a substantial part of their R&D resources to develop product innovations and why a majority of empirically observed innovations tends to be product rather than process innovations (Pavitt, 1984).

Our answers to this riddle is that most markets are organized markets rather than pure markets (Lundvall, 1985; 1988). What appears as a market is actually an institutional mix between elements of organization (dominance, loyalty, flows of qualitative information, cooperation etc.) and the traditional elements of the market (commodity, price, sellers, buyers, and acts of exchange).

Especially in markets where needs and technical opportunities are in a state of constant flux the organizational elements will be strong. Only in markets for standardized products with stable use value characteristics will anonymous markets be adequate. Technical change has a contradictory effect upon markets. In so far as it is expressed in new use value characteristics it strengthens the element of organization in the markets. To the extent that it establishes standards and homogenous quality it reduces the element of organization.

It comes as no surprise to engineers and marketing managers that user-producer relationships are organized and often involve long term cooperation. What to the theoretical economist appears as a pure market and to the lawyers appear as an arena for cheating and haggling will often appear as a framework for informal cooperation to engineers and people engaged in marketing.

But why should firms prefer organized markets to vertical integration? One reason is the flexibility offered by the organized market. The formal hierarchy might be more efficient in transmitting and processing information reflecting learning and investment in firm-specific codes and channels of information but it is not a very flexible way of organizing economic activities.

When the environment changes rapidly, and radically, the old inter-firm network becomes inapt for solving the problems encountered. And here the organized market offers a more flexible solution. Even if informal user-producer cooperation is long term it is not necessarily for a lifetime.

The second argument refers to interactive learning. Producers, interacting closely with competent, advanced, and demanding users, will be involved in a process of learning-by-interacting crucial for the success of product development. A formal integration would, however, exclude both parties from interacting with other users and producers. Such a narrowing of the learning process would result in a gradual erosion of the competence of both parties.

Cooperation as a Dimension of User-Producer Interaction

Not all user-producer relationships involve cooperation in any meaningful sense of the word. Most consumer markets are organized unilaterally by the producers and qualitative information is gathered through market analysis and diffused through advertising. When users are professionals organisations cooperation becomes more frequent, however. A considerable share of innovations involve cooperation with customers and subcontractors at several different stages of the process.

Cooperation between users and producers might be asymmetrical in terms of power relationships and come very close to the relationships in a formal hierarchy. A large firm may, for instance, specify in detail what a subcontractor should do, including the price to be charged. If the subcontractor does not fulfil the expectations the large firm may exclude him/her from further business or choose to integrate the small firm. This kind of absolute dominance will increase the flexibility of the large firm but it does not contribute to its innovative capability. For the large firm this kind of subcontracting may be regarded as a defensive response to uncertainty.[2]

Cooperation between users and producers where both parties take active part in developing the innovation may be regarded as a process of interactive learning and as a positive sum game where both parties may benefit. However, cooperation involving interactive learning raises many problems reflecting the uncertainty inherent in a process of innovation and the asymmetrical distribution of information between the two parties. The users have to expose their specific competence. There is plenty of room, for each party, to abuse this kind of sensitive information. If both assume the other to be opportunistic, cooperation may very well be highly cumbersome.

Even in such a situation the small firm may benefit from interaction. Its organization may be upgraded in order to meet far-reaching requests from the dominating user.

We have discussed above why the organized market may be preferred to vertical integration; but how can cooperation for innovation survive in spite of the uncertainties and asymmetries involved? Two different kinds of explanations may be considered. The first assumes rational behaviour as it appears to prevail in mainstream economics. We may argue that it is "rational" to be loyal to your partners because trust becomes crucial to success. Agents who get a bad reputation in terms of breaking loyalties may be excluded from processes of interactive learning. This kind of explanation presents interfirm cooperation as kept together exclusively by rationality. All parties cooperate in so far as it can be assumed to be in their own interests.

Another explanation assumes that individuals and organisations act from more complex motives. At the individual level the agents may be endowed with a moral dimension telling them that cheating is bad and they might feel more at ease in an atmosphere of mutual trust than in one of cut-throat competition and struggle (Etzioni, 1988). At the social level institutions, through routines and codes of conduct, put strict limits

2 Even in such a situation the small firm may benefit from interaction. Its organization may be upgraded in order to meet far-reaching requests from the dominating user.

to what is regarded as acceptable conduct from organisations and individuals.

We believe that actual cooperation reflects both kinds of elements. Were individuals and firms to join and leave alliances exclusively on the basis of rational calculations cooperation would become extremely difficult. Non-economic dimensions must support cooperation– a feature which has several important implications.

First, it emphasizes the importance of the social, cultural,and, even, moral environment of cooperation. Different national settings will, for instance, form environments more or less favourable to cooperation. Another consequence is that the frequency of registered formal agreements may be misleading as an indicator of the importance of actual cooperation in a given economy. A high frequency of formal judical agreements may reflect expected non-cooperative and opportunistic behaviour; formal contracts may be regarded as attempts to counter opportunism. A low frequency of registered cooperation may also reflect that informal cooperation makes formal agreements unnecessary.

Some scholars studying the characteristics of the Japanese system of innovation (Dore, 1987) have pointed to the cultural roots of the close cooperation between users and producers. Most experts on Japan would agree that cooperation between firms is more developed there than in the US and Europe (Imai & Baba, 1989; Freeman, 1987). But data banks of registered interfirm agreements, within Japan, will often come out with a low frequency (Hagendoorn & Schakenraad, 1989).[3]

In quite another context, Håkansson & Johanson (1988) refer to a Swedish study of cooperation between firms in technical development in which more than two-thirds of the cooperative relations were informal. However, empirical evidence on the relative importance of formal versus informal cooperation appear to be only sparse at the present stage. Here we see a need for further research at the European level aimed at understanding international differences in the sociocultural preconditions for cooperation.

3 The authors point to lack of information as an explanation of the low frequency of observed interfirm agreements (4%intra-Japanese vs. 27% intra-USA agreements). We may propose that another part of the explanation could be the less formal manner of cooperation in Japan.

From User-Producer Interaction to Networks

User-producer interaction is obviously only one kind of interfirm cooperation and in studies of interfirm agreements it tends to play a secondary role. More important roles are given to joint ventures and specific technology-exchange contracts. Here we shall argue that the way markets for producer goods are organized is important not only in its own right but also indirectly. An economy in which firms often take part in vertical interfirm cooperation will be better prepared to develop other kinds of informal cooperation than one where firms operate at arms' length from their users and producers. User-producer links may be regarded as organically grown relationships which may be supplemented by horizontal or diversifying relationships.

Taking all kinds of cooperation into account brings us to the concept of networks. A network represents durable non-market relationships between a cluster of firms operating in different parts of the economy. User-producer relationships represent one important set of relationships in networks but other sets involve horizontal and diversifying relationships.

Cooperation and Uncertainty

It is increasingly realized that the present period is one of transition from one kind of economy to another. This transition hypothesis has been given many names in the last decades (the information society, the post-industrial society, the post-fordist society, etc.) but most of them refer in one way of another to a change from standardized mass production to a more flexible and differentiated production system.

The fact that we are in a period of transition will in itself increase the uncertainty in the economy reflected both in uncertain market expectations and in uncertain efforts to develop and introduce new technology. In such a period it is important for firms not to get locked in in terms of products, process technology, markets and organization. Flexibility through cooperation might be regarded as one way to keep many possible roads open. Viewed in this perspective the search for flexibility is a transitory phenomenon, involving costs of transition reflected in slow growth and unemployment.

Another recent interpretation with quite radical implications has been put forward by Imai and Baba (1989 op. cit.). According to their analysis several new features point towards interfirm cooperation and networking as an institutional mode of increasing importance. The rate of technical change and a shortening of the product life cycle increase the importance of continuous innovation. The systemic character of innovations, reflecting the need to combine different generic technologies into new products, makes it increasingly necessary to combine the competence of different firms. And the increasing importance of combining competence in hardware, software and services works in the same direction.

According to this analysis the role of entrepreneurship is changing. The capability to build and interact in networks becomes of crucial importance and in the long run some of the classical functions of the individual entrepreneur will become embodied in networks. According to Imai and Baba, Japanese competitiveness reflects the fact that domestic firms have developed a highly sophisticated structure of networks and cooperation, stimulating interactive learning and innovation.

We consider this hypothesis to be highly relevant. If correct, it has far-reaching implications for the economic system. It also implies that habits of thought of managers and policy-makers should be revised. The cultural and social dimension embedding the economy becomes of increasing importance.

Small Firms in Small Countries – Different Perspectives for Focus on Interfirm Cooperation

Cooperation between small firms from small countries should not be considered as yet another magic formula, capable of solving all sorts of economic problems in these countries. However, it appears as if a growing amount of evidence points at the importance of strengthened interfirm cooperation as a major strategic tool for thesurvival of many of these firms.

To bring this discussion a bit further, classification appears to be helpful. It is essential to keep in mind some differences within the small country group. In terms of *commodity* exports, most small high income OECD countries are characterized[4] by a high degree of international specialization in natural resource-based products – e.g. Sweden, Norway and Finland in wood manufacturers, paper & pulp, metals; Denmark, Belgium and the Netherlands in agricultural products. This "profile" of exports coincides with specialization in products with low or medium

4 For a comparative study of intenational export specialization of 23 OECD countries 1961-85, see Dalum (1989).

technology intensity (more or less equal to low or medium growth products).

Switzerland and Ireland turn out as exceptions from the above pattern. They are highly specialized in chemicals, Switzerland also in engineering. In terms of technology intensity both countries are specialized in high tech products (more or less equal to high growth products); Ireland, only so from the late 1970's.

Another classification variable is *company structure*. Some countries like Sweden, the Netherlands and, to an increasing degree, Finland have somehow compensated for this apparently unfavourable specialization pattern by being home nations for some large multinational enterprises (MNEs) pursuing increasingly global strategies, e.g. Ericsson, Electrolux, Volvo, Asea Brown Boveri (Sweden/Switzerland); Philips, Akzo, Unilever & Shell (the Netherlands/UK); Nokia, Kemira, Neste (Finland). Others, like Denmark and Norway (with Norsk Hydro as an exception), have not emerged in this category of "home" nations.

According to the company variable, major differences emerge between Switzerland and Ireland. Like Sweden, Switzerland is a home nation for a certain amount of large MNE's especially in chemicals, pharmaceuticals and some segments of engineering. Ireland, on the other hand, appears to be the archetype of a "host" nation for MNE assembly plants (computers, semiconductors) or chemical processing plants.

The context for policies inducing interfirm cooperation is quite different between these different groups of small countries. In (non-EC) countries like Sweden and Switzerland it is obvious that the general "health of the nation" is tightly correlated to the performance of, say, 10-20 domestically based MNEs. In these countries the large firms tend to be the ones organizing and structuring networks of cooperation involving many of the small and medium-sized domestic firms. Here, the policy context for inducement of innovative activity in small firms is more one of the "classical" discussion of how to create symbiotic relationships between small and large firms.

In a country like Denmark, dominated by small and medium-sized firms, the context is quite different. Given the trend towards more systemic technologies and the acceleration of the internationalization process (also spurred on by "Project 1992"), it is no surprise that increasing concern has been concentrated on how to create more dynamism in the small and medium-sized enterprise sector (the "SME sector") – simply because it comprises the major share of the manufacturing sector.

From the same perspective it would be no surprise if the Irish authorities were concerned about how to spur new dynamism into its indigenous SME sector to create some sort of "lever" vis-a-vis the foreign MNEs. This concern may be relevant also in the case of the lower income small EC members, Greece and Portugal.

Some Issues on Interfirm Cooperation from the Danish Case

During the 1960's and the first part of the 1970s it was quite common to praise the flexibility of the Danish SME sector as maybe *the* most important factor behind the very high growth rates of production and productivity. A sometimes romantic perspective on the small firm culture, paired with a rather liberalistic industrial policy, predominated. It is true that it looked as if it was easier for the Danish production sector to adjust to the economic aftermath of the First Oil Crisis economic context. The painful adjustment process experienced in many heavy industries and in traditional energy intensive sectors (e.g. steel) in countries like Sweden and Belgium was avoided – simply because these sectors were nonexistent in Denmark.

However, during the late 1970s and more so in the 1980s, in the face of weak performance in terms of employment, balance of payments and economic growth, the dream has broken down. In recent years the major issue has been the lack of "industrial locomotives"; and how to create such "locomotives" has dominated the general discussion of the industrial future of the country. This debate has been spurred on by the vision behind the creation of "Project 1992" – dynamic restructuring of European industry by reaping the benefits of specialization/economies of scale, brought about by intra-European mergers and acquisitions. Thus a virtual wave of mergers & acquisitions has been the major feature for example of the dominant food processing and agricultural machinery industries in Denmark.

On the other hand, the traditional Danish SME-dream has not been totally forgotten. The government has tried to solve this inherent dilemma by launching two parallel industrial policy programmes in 1989. One dominated by support schemes for collaborative R&D projects between large firms; the other supports the education and authorization of "network brokers" (persons with the task of creating networks between small firms) and the initial establishment of such networks.

Whether or not induced by this latter programme, a substantial amount of network cooperation has, recently, been established between small firms, especially in the furniture and textile industrieso concentrating mainly, on cooperation in marketing abroad.

However, this (at least potential) split between intra-large firms and intra-small firms cooperation may prove to be problematic. Some of the most fertile industrial cooperation projects may very well involve cooperation between large and small firms as well as government bodies/public utilities/public research institutes – i.e. quite complicated network structures by means of which competitive advantages in private production *and* government regulation may be jointly exploited.

To support this perspective, we may refer to the electronics industry. Taking our point of departure in Table 1 below (c.f. Dalum et al 1988; 1989) we have found that the strength of the Danish electronics industry is concentrated in small scale markets (instruments amounted, thus, to around 40 per cent of Danish electronics exports in 1985). Firms belonging to the "global" category export more than 90 per cent of production, while the firms producing "tailor-made" instruments export 40 to 50 per cent of their production. With the exception of two firms (Bruel & Kjaer, around 2.000 employees and Radiometer, approximately 1.000) all the firms in these areas employ less then 500-600 persons.

	Industry	
Scale	Multidomestic	Global
Mass markets	Telecommunications (moving towards global)	Semiconductors Computers Consumer electronics
Small markets	Tailor-made instruments	Specialised instruments

Table 1: Classification of electronic industries
(Source: Dalum, Fagerberg and Jørgenson, 1988)

Most of the successful products and firms are "classical" examples of intensive and successful user-producer interaction *combined* with the existence of a strong domestic research environment in a given, fairly narrow field. The most obvious examples are to be found in interaction between hospitals, university research and electromedical firms.

However, it is only the "global" firms which have been established as members of exclusive "clubs" of world-wide leaders in their respective oligopolistic niche markets. And even in these cases they often compete, directly with large international firms, such as Hewlett Packard, Tektronix, the electromedical division of Siemens, etc. Recently, some of the Danish entrepreneurs (often the founders or their relatives) have argued for a merger of some of the major Danish "global" electromedical firms to avoid the high cost of maintaining parallel worldwide service and sales networks for example.

To spur the internalization process in the "tailor made" instrument firms, other sorts of networks appear to be more relevant. The firms often have often tight cooperative relations with public utilities (control instruments for production and distribution of electricity and/or heat) and big private domestic firms (e.g. in food processing and brewing). Support for cooperation exclusively between small firms will be of little help in these cases.

Some Tentative Conclusions

In the first part of this contribution to the discussion on the Cooperation Phenomenon we argued that formal cooperation is only the tip of an iceberg reflecting cooperation between users and producers as well as the swapping of knowledge between competing firms. It was also argued that cooperation involves "non-economic" variables which differ between national systems of innovation.[5] They will reflect a specific historically based sociocultural context. This implies that national and transnational policies aiming at exploiting the potentialities of interfirm cooperation involving small firms must take into account such international differences. In this context we emphasized the need for systematic research at the European level on the relative importance of differences in the sociocultural context for cooperation and competition.

Other national differences are easier to observe and analyze. Differences in production and export structure and especially in the role of large firms in the respective economies should also be reflected in policy strategies. Through examples from the Danish economy we illustrated that even within one single sector, electronics, the potentialities of inter-

5 For a discussion of the concept of "national systems of innovation", see Andersen & Lundvall (1988).

firm small firm cooperation should reflect the specific conditions of subsectors and product groups.

The general conclusion from this paper is that policy-making in this area must be well informed and based upon a thorough knowledge of the specific character of national systems and of sectors and subsectors. This might put some limits to the efficiency of general and transnational policy making at the European level. in order to be efficient such a policy must interact closely with national units with an analytical insight into the specificities of the national systems of innovation involved.

References

Anderson, E.S. and Lundvall, B.-Å. "Small National Systems of Innovation Facing Technological Revolutions - an Interpretative Framework" in Freeman, C. and Lundvall, B. -Å. (eds.) *Small Countries Facing the Technological Revolution*. London: Pinter Publishers, 1989

Contractor, F.J. and Lorange, P. eds. *Cooperative Strategies in International Business*. Massachusetts: Lexington Books, 1988

Dalum, B. "Export specialization, competitiveness and national systems of innovation", *Institute of Production*, Aalborg University Center, mimeo, 1989

Dalum B.; Fagerberg, J. and Jørgenson, U. "Small open economies in the world market for electronics: The case of the Nordic countries" in Freeman, C. and Lundvall, B. -Å. (eds.) *Small Countries Facing the Technological Revolution*. London: Pinter Publishers, 1988.

Dalum, B. ; Fagerberg, J. and Jørgenson, U. "The Nordic electronic industries facing the world markets", Draft version of a report to the Norwegian technology history project. Institute of Production, Aalborg University Centre, 1989

Dertouzos, M. -L.; Lester; R. K.; Solow, R.M. and the MIT-Commission on industrial productivity. *Made in America*. Cambridge, Massachusetts: The MIT Press, 1989.

Dore, R. *Taking Japan Seriously*. London: The Athlone Press, 1987.

Etzioni, A. *The Moral Dimension*. New York: Collier-MacMillan, 1988.

Freeman, C. *Technology and Economic Performance: Lessons from Japan*. London: Pinter Publishers, 1987.

Hagedoorn, J. and Schakenraad, J. "Strategic Partnering and Technological Cooperation" in Dankbaar, B. et al. (eds.) *Perspectives in Industrial Economics*. Kluwer Academic Publishers, forthcoming.

Haywood, B. and Bessant, J. *The Swedish Approach to the Use of Flexible Manufacturing Systems*. The Innovation Research Group. Brighton Polytechnic, 1987.

Håkansson, H. and Johanson, J. *Formal and Informal Cooperation Strategies in International Networks* in Contractor, F.J. and Lorange, P. (eds.), 1988.

Imai, K. -I. and Baba, Y. "Systemic Innovation and the Cross-Border Networks", paper presented at the International Seminar on Science, Technology and Economic Growth, June 1989, Paris, OECD.

Katzenstein, P. *Small States in World Markets*. Ithaca: Cornell University Press, 1985.

Lundvall, B. -Å. *Product Innovation and User-Producer Interaction* Aalborg: Aalborg University Press, 1985.

Lundvall, B. -Å. "Innovation as an Interactive Process – from User-Producer Interaction to National Systems of Innovation" in Dosi, G. et al. (eds.) *Technology and Economic Theory*. London: Pinter Publishers, 1988.

Pavitt, K. "Sectoral Patterns of Technical Change: towards a Theory and a Taxonomy", *Research policy*, Vol. 13.

Williamson, O.E. *Markets and Hierarchies: Analysis and Anti-trust Implications*. New York: The Free Press, 1975.

Cooperation in the European I.T. Industry – Can the Smaller Countries Benefit?

P. Murtagh

Introduction

This paper highlights the importance of the common intellectual role of smaller countries in the I.T. arena i.e. the ability to carry out quality basic research and applied research and development which form a solid basis for future industries. Such countries need encouragement to carry through the kind of developments necessary if their industries are to emerge as industrial peers within the European Community. The paper describes how industrial I.T. cooperative programmes are helping to achieve this goal and suggests how future policies at both a European Community and National Government level should be tailored to build on the results achieved.

Information Technology is associated with the following areas:

- Information systems, including software and services
- Electronic components
- Industrial automotive systems
- Computer integrated manufacturing
- Professional electronics
- Consumer electronics
- Office equipment
- Telecommunications

Accurate statistical data on the above is difficult to pin down due to severe problems in its collection and definition. However, we can make the global observation that I.T. industrial output represents 4.4% of European GDP. By 1993 it is expected that this will grow to 6.7% which will be more than any other industrial sector. Moreover, nearly two thirds of the industrial and services sector are dependant on I.T. for their efficiency and competitiveness.

IT therefore represents a major factor in the creation of wealth and consequently increases in the standard of living. Competitiveness, which can be seen as mirroring of national capacities i.e. resources, skills etc., is heavily dependent on the appropriate application of this technology.

The low infrastructural requirements (particularly in software) of the industry has given a role to the smaller EC countries in the development and use of I.T. Many of the organizations involved are SMEs. Such environments, particularly where competitive pressures are high, favour short term returns with usually a consequentially low level of innovation and short product life cycles. These conditions are not conducive to high R&D investment and there is a tendency to be trapped in a survivalist mode of operation where planning beyond the short term is not feasible. Such an industrial fabric is very fragile with a greater susceptibility to reverse technology transfer in technological/scientific cooperative ventures, open to the brain-draining effects of an increased competitive demand for I.T. skills, can be influenced by extra-European interests and faces increased competitive pressure from extra-national industries.

If smaller countries are to benefit, it is important that their industries move from a startup phase to one where there is a more mature and longterm outlook and where skill development and retention will be a key aspect of their survival.

To illustrate these points an outline of collaborative EC programmes in IT and telecommunications (ESPRIT & RACE) and an assessment of how they are helping to increase the level of knowledge and skills in smaller countries, based on recent studies, is given below.

Major Characteristics of the Six Smaller EC Countries

The twelve EC member countries differ in terms of size, skills, national income, technological level, rate of growth and degree of dependency on other countries, i.e. import/export profiles.

Two major blocks can be identified – the larger and the smaller countries. In this definition the smaller countries are identified by their relatively low population, by area and by national income (i.e. population less than 10 million, area less than 130,000 sq.Km.) Table 1 indicates the relative technological development for each small country.

INDICATORS	B	DK	GR	IRL	NL	P
Population a) (1,000- Hab.) 1985	9858,3	5113,7	9934,3	3540,0	14991,6	10157,0
Geographical area a) (Square km)	30,5	43,1	132,0	70,3	41,8	92,4
GDP BECUs 1986	114,4	84,1	40,5	24,9	170,0	28,9
Exports (goods) b)	64,9	29,5	13,9	56,9	54,8	29,3
Technological level						
GERD (%GDP) c) 1987	1,59p	1,37c	0,33 (1986)	0,96p	2,22 (1986)	0,45 (1986)
Total R&D Personnel - FTE 1987		22000	6,091 (1983)	6343 (1986)	62830 (1986)	10570 (1984)
Business Enterprise expenditure (million $) 1987	1362,8	542,2	58,6 (1986)	148,6p	2469,3c	67,41 (1986)
Technological Balance of payments (technology coverage ratio) 1985	0,45	1,14	-	2,19 (1983)	0,44	0,11

Table 1: Major characteristics of the 6 smaller EC countries

Sources:
a) Eurastat
b) OECD, L'Observateur de L'OCDE
c) OECD, STIID, DATA BANK, April 1985 (p - provisional, c - national estimated)

Even within this group the countries do not have the same level of development and technological capacity.

With the exception of the Netherlands (which has one of the twelve major I.T. companies) 21st in the 100 leading worldwide data processing companies they are not major players in the I.T. sector either in terms of R&D and production. Table 2 describes the relative level of production capability for each country.

Sources:

INDICATIONS	B	DK	GR	IRL	NL	P
Production a) (MUS$)-	2360	1083	-	2409	4940	
Export a)	1798	930	-	2711	4667	456
Import a)	2460	1324	-	1767	5396	467
Production a) (%GDP)	3.0	1.9	-	13.6	3.9	-
b) E/I ratio - office made computers	0,64	0,30	0,01	2,26	0,67	0,20
E/I ratio - comms electromics	0,83	0,84	0,10	1,62	1,09	0,44
IT total R&D personnel FTE (electronics, office machine communications)	3725	1029	-	633,6	-	-

Table 2: Major characteristics of the 6 smaller EC countries

a) Mackintosh yearbook, electronics data 1987
b) OECD - Foreign trade data bank

Collaborative R & D Programmes in IT

Collaborative R & D programmes in IT in Europe started in the early 1980s with the UK Alvey Programme which was a national initiative in response to the Japanese 5th Generation Programme. Other major countries also had their own collaborative initiatives mainly to provide stronger links between universities and industry.

There was little or no evidence of collaborative programmes in the 6 smaller countriesat that time and across Europe as a whole, collaboration between firms from different countriesin was nearly non-existant.

From 1984, at a European level, three major programmes were formally established in high technology fields: two EC programmes – ESPRIT and RACE and the EUREKA Programme which has a major involvement in I.T.

The cooperation was geared towards encouraging a sharing of resources, results and risks. It turned out that greater cross fertilization / exchange of knowledge created the necessary leveraging for research resources to allow the delivery of more ambitious results more quickly and sometimes at a lower cost.

The success of this R&D-based technological cooperation has facilitated a possible bridging towards commercial cooperation and joint ventures with the hope of new products at lower cost being introduced into the marketplace.

The involvement of smaller countries with a weak IT base in such R & D collaborative programmes has had a number of implications both beneficial and detrimental.

Benefits

- Diffusion of technology.
- Knowledge exchange.
- Increased visibility in the market and recognition of technological competence.
- Proximity of suppliers.
- Learning and getting acquainted with the new technology developments (catching-up).
- Training the management of small companies to handle big projects in an international environment. The experience gained gives the smaller company the confidence and capability to handle projects of a similar nature both nationally and internationally. This experience might have been impossible to obtain in a typical small country environment.
- Being part of a larger network of cooperating organizations.

Dangers/Threats

- The I.T. base of many of these smaller countries is still very weak. Much of the expertise resident in these countries is still based in universities and research institutes which have weak connections with local industry. The result is that the original work of these countries is taken advantage of more easily by the richer and more experienced cooperative partners. The risk of reverse technology transfer is therefore high if the I.T. base of that smaller country cannot immediately and directly avail of research results.
- The I.T. base is mainly composed by of SMEs who tend to have a lower ability to utilize research findings quickly and widely.
- Vertical cooperation with competitors of national companies can dilute the national content of a product with the result that the smaller countries' value added contribution may tend to be the least profitable.

- Dangers of takeover and acquisition of the most promising companies before they reach their mature stage.
- To maximize involvement in such programmes, the SMEs of small economies tend to work on areas of major interest to the Big 12/large European companies rather than on those of more importance to indigenous companies.
- Problems in protecting background information.
- Patent and copyright systems are imperfect.
- Overheads of cooperation, costs of participation, travel, communications for the more remote countries are high.
- Benefits obtained are uneven and can be used by the larger and better equipped organizations to greater effect.
- Orientation/mobility of skills and human resources can be altered to the detriment of a smaller economy.
- Relatively few alliances with companies of small countries have taken place. Where they have occurred, they have tended towards takeovers.

Networks and Diffusion of Technology

While there are a number of contentious issues arising out of cooperation particularly for smaller countries, it is here to stay for the foreseeable future. Benefits exist and it is up to national governments to maximize their effects.

Some key points of awareness need to be addressed:

- national capacities:

  There is a tendency towards not using local SMEs in small countries where there is a general belief that capabilities and skills which can contribute to the infrastructure of an economy are not available. (i.e. small companies can only do small jobs). These cooperative programmes have consistently demonstrated that it is untrue.

- role in a European Context

  Small countries need to identify differentiators which highlight their role in a trans-European environment. Where do they contribute and how do they fit in?

The main approaches for national government include:

- role of programmes:

Small member states should play a more active role in the definition and design of European programmes. This would maximize the relevance of programme outputs for small countries and help them to participate more effectively in a European context. Too often there is a tendency to accept secondary roles in large scale programmes whereas well defined primary roles should be the aim.

- funding:

  Funding should be tailored to maximize the benefits of cooperative R&D ventures particularly where results are achieved more quickly and require to be taken advantage of in a shorter timeframe.

- facilitate the participation of their own national companies.

  There is a great need to increase the integration of small companies into national organizations to help to focus skills on indigenous issues and provide a fertile starting point for promising SME spin-offs?

- better diffusion of information:

  The general level of awareness of the achievements and results among SMEs not directly involved is still low. There is a need to diffuse information more effectively. One method could be through the funding of technology users clubs/workshops.

- facilitators:

  Smaller countries can maximize the benefits from their participation in these types of programmes, taking advantage of "facilitators" who can have an important role in:

  – in the diffusion of technology.
  – in the identification of opportunities.
  – in building consortia
  – in maximizing the use of skills and human resources available.

As stressed by a number of contributors to this book such intermediaries should have the necessary international experience to assist both in the formation of policy and strategy and their implementation which would help maximize the return from cooperation, particularly in the I.T. and high technology sectors.

Cooperation – The Small Firm, Small Economy Experience[1]

D. O'Doherty

This paper takes as its starting point one of the major issues highlighted by Francois Chesnais of the Organization for Economic Cooperation and Development (OECD) in his seminal review of the growth and significance of the internationalization and cooperation phenomenon[2]. He concluded that "firms of the smaller industrialised member states, and of countries with weaker technology bases, may have a particular interest in, and face particular constraints with respect to international technological cooperation". P. 112. We know that these are also the economies which are dominated by small firms and which have generally weak industrial and technological structures.

Chesnais went on to say that "such cooperation has so far been undertaken mainly through partnerships with large firms from large (or relatively large) countries and through participation in alliances led by such firms and countries". He noted both the lack of actual experience and also of data and studies on cooperation between the smaller countries/smaller firms themselves or of small firms in association with larger firms from larger countries.

We can all concur on the latter point, just as there are too few studies on the internationalization and cooperation phenomenon generally. It is also hard to disagree with the general thrust of his substantive points about the smaller countries being left out of, or at least having major problems in being part of, the current process of the internationalization of science and technology and the strategic alliances that go with it. And it is certainly true that it is the smaller firms and smaller economies which most need the complementary assets and capabilities potentially provided by partnerships and alliances.

1 This paper draws on material prepared for the 6 Countries Programme on Innovation Workshop "Strategic Alliances in R & D" Ottawa Oct 1989

2 Chesnais F: "Technical Cooperation Agreements between Firms" STI Review No. 4. OECD, Paris December 1988"

"R and D" – too narrow an approach?

In Chesnais' paper however there may have been an implicit overemphasis on technology generation. Such a perfection militates against small firms and small countries themselves making the correct strategic moves to enable them to become more involved in, and to gain the greatest benefit from, the internationalization process. It also militates against larger firms and countries understanding how they can assist the smaller entities in that process by avoiding what could be a zero or negative sum game for small firms. paradoxically, this could ultimately damage their own longer-term interests given the innovation potential of small firms.

It is potentially positive sum game of conceptualising internationalization as affecting the whole innovation process, involving technology application, transfer and diffusion – where new products and processes are actually produced – rather than only the internationalization of R and D for technology generation, which is the key issue for small firms and countries.

For the larger firms and countries these have usually meant much the same thing in practice. Large corporations have long undertaken industrial R & D programmes, often with "national" support, and with little difficulty then commercialised and marketed the results in their established markets. Their approach may have tended to leave out the increasingly important role of the "user as innovator", for example, but this relatively problem-free, almost linear, innovation process has been the norm for many large enterprises. Technological and market changes have however more recently dictated a change in strategy and approach. This now means that even the largest and most successful corporations (usually originating in either the US or Japan) engage in strategic partnering and collaboration – normally with each other and most visibly in research for technology generation, which for them carries the highest level of risk and uncertainty. But the significant thing is that they are collaborating in such R & D with a view to the commercialization and marketing of the "outputs" of that research, in both their own and their partners' captive markets.

The high profile and apparent relative significance of "technology-generating" research for the very largest firms, where it has in the past been a proxy for innovation, has led many analysts, public servants and even smaller firms to focus almost exclusively on the generation of new technology as the only important activity in the overall internationalization of technology. The President of Olivetti has highlighted the limitations imposed by this attitude on European companies, countries and research institutions[3], which he sees as exacerbating the technology gaps between Europe and the other members of the Triad. Its larger more that its smaller entities perhaps are failing to bring their own research to fruition and to engage strategically as technology "leaders" in interfirm alliances with Japanese and American counterparts aimed at bringing new products and processes to the marketplace. Thus, despite the fact that European research expenditure is twice that of Japan and two-thirds of the US figure the record for commercialising new products and processes in the advanced technologies in Europe is rather poor.

This philosophy and approach can also be observed in the programmes of the European Community – particularly in the "Framework Programme for Research" (3rd programme budget: 5.8 billion ECU 1990 - 1994) where the term "precompetitive research" reinforces the point that it is largely focusing on the research and technology generation aspects of the innovation process. It should be noted here though that the collaborative element which is prescriptive in all EC research programmes appears to be significant for those SMEs which have become involved in such programmes – in building project-specific alliances but particularly in generating valuable contacts and longer-term cooperative networks.

A "total innovation" approach

How realistic then is it to extend the concept of "internationaliZation in science and technology" to encompass the whole of the innovation process – including technology generation, but also application, transfer and diffusion? And what are the implications of such an extension for our understanding of what are strategic alliances and the role of small firms and small economies? It has already been argued that the largest and most successful companies have no particular problems with an "innovation" approach. For them it is the norm. Within such an approach the concept of collaborative R & D itself is expanded to encompass a number of situations in addition to technology generation e.g. in specific

3 de Benedetti Carlo: "Europe's New Role in a Global Market in "A High Technology Gap", Council on Foreign Relations, New Yor 1987

technology "applications" and the adaption of transferred technology. Side by side with the increase in research collaboration which has led to this Workshop there has been significant growth in the other forms of technology agreements and exchanges. In the OECD publication already quoted the results of a survey carried out in the mid 1980s on licensing and the other technology transfer activities of firms world-wide are reported.[4]

They indicate an overall increase in licensing-out during the 1980s (Charts 1 and 2, Source Vickery Graham, " A Survey of International Technology Licensing") with 60 percent of respondents reporting increased activities but more particularly major growth in joint ventures, (reported by 81 per cent of enterprises) and technical service agreements, clearly involving continuing relationships. the largest and smallest firms were particularly active in all aspects of technology transfer, with medium – large companies somewhat less so. In the same publication technology exports by Spanish companies are analyzed, showing a major increase in activity in and from that country.

A recent study of patent licensees in Ireland[5], relating to data collected in 1986, indicates that while the level of activity was quite low, many of the agreements, and the most successful, involved a continuing relationship between licensor and licensee, including exchanges of personnel and knowhow. The significance of this is highlighted by the fact that the OECD technology transfer study indicates that over 90 percent of enterprises surveyed are licensing out technology which they are actively using themselves. There appears to be conclusive enough evidence to show that an increasing amount of technology sharing is going on between firms in various types of relationships with each other, not just in collaborative R&D but also arm's length licensing and increasingly joint ventures, as well as the more classical intra-firm transfers of multinational enterprises.

4 Vickery Graham; "A Survey of International Technology Licensing" STI Review No. 4 OECD, Paris, December 1988

5 Allen T.J. Cogan D.J.., O'Doherty D.., O'Sullivan B.A. "Direct Acquisition of Technology by Irish Industry" Science Policy Research Centre, Dublin, October 1989

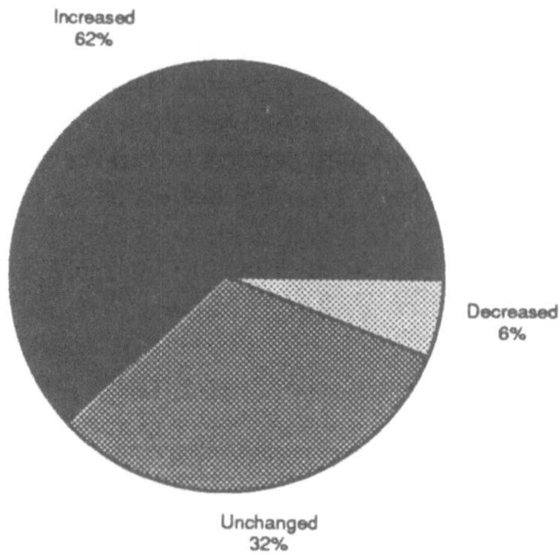

Chart 1: International Technology Transfer Survey
Licensing Activities of Firms (%)

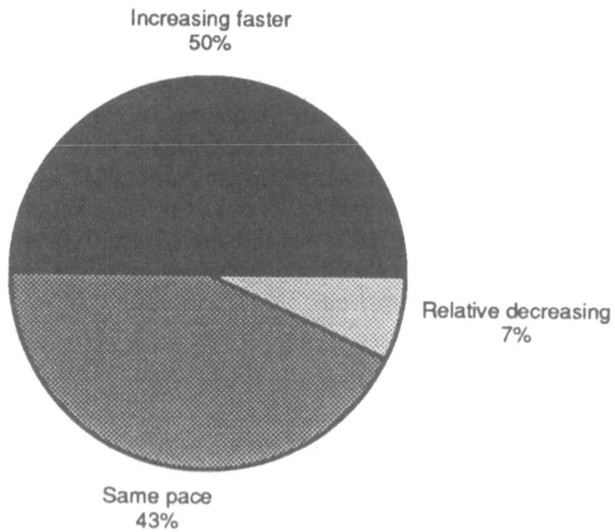

Chart 1: International Technology Transfer Survey
Other T.T. Ativities of Firms (%)

Strategic Alliances for Innovation?

It is obviously a matter of definition but these relationships appear sufficiently close and interactive to be called alliances and partnerships. Whether they are "strategic" or not depends very much on where the individual firm or country's major capabilities and requirements lie on the technology generation to diffusion spectrum. In the case of smaller firms and smaller countries, these are more likely to be at the latter end, so for them transfer is or should be the "strategic" approach.

On the assumption that strategic alliances can and do take place from generation through to diffusion and the R&D does not have to be confined to the generation stage only, what is the significance of this for small firms and small economies?

As well as encompassing a much wider range of possible relationships for and between individual firms, this perspective highlights the need to look at all the actors and potential partners in the internationalization process. These include:

- firms, operating individually or collectively
- universities
- public research organizations
- public and private intermediaries
- the state itself, operating both nationally and supranationally (e.g. the European Community)

The relative positions of these actors in international research collaboration in one small country with a preponderance of small firms (Chart 3, Source: Census of Industrial Production) is illustrated by the latest figures for Irish participation in the research programmes run by Directorate General XII – Science, Research and Development of the European Commission.

Between 1983 and 1988 inclusive, the contractors involved were as illustrated in Chart 4 (Source: CEC). Around 60 percent of the collaborative projects went to the universities, representing only 20 percent of the organizations involved in such partnerships, with industry providing nearly 55 percent of the contractors but being involved less than 12 percent of the projects. In the vast majority of cases the collaborative arrangements, although mixed in terms of the types of partners involved, were almost exclusively with overseas-based partners, There was little university-industry interaction in joint projects at "local" level.

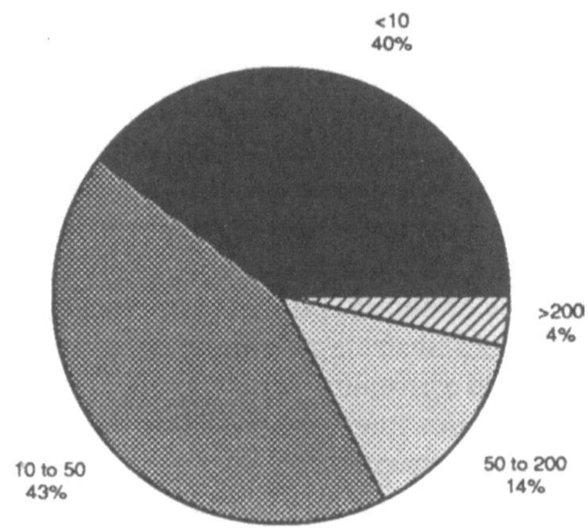

Chart 3: Scale/Structure of Irish Industry
Size of firms involved in international research collaboration

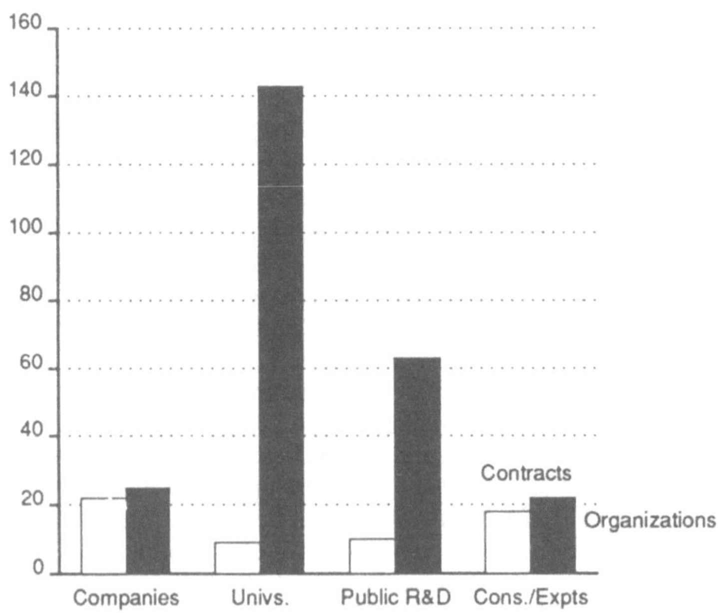

Chart 4: Projects involving Partnership

The imbalance is not so great in the programmes of DGXIII – Information Technology, which include the "applications" aspects of ESPRIT and other I.T. – related programmes, in addition to technology generation-type activities. But both the absolute position of firms and their position relative to the universities raises many issues for all the actors involved, particularly the state.

In most cases the involvement of the universities in these projects and alliances can hardly be called "strategic" in either a national or industrial sense. Individual researchers become involved because project funds are available in attractive research areas. But little or no account is taken of the underlying objectives of the individual projects in commercial and economic terms and the fact that opportunity costs may arise through the diversion of resources which may come directly from university sources but are ultimately funded by the national government. Universities as institutions may themselves be losing out in partnerships with large overseas organizations and failing to adequately asset their intellectual property and commercial rights in the results of their research. However, as primarily educational institutions they find it particularly difficult to respond to such problems. This illustrates the need for governments to consider more carefully their policies and approaches towards funding such institutions given their level of perhaps unconscious incorporation into the international level if, for example, the benefits to the rest of the European Community outweigh those, if any, to the smaller member state in particular.

For industry and consequently industrial and technological policies the lessons are even more clear-cut. Even if the 24 companies quoted above are gaining from the contracts and cooperation involved and are generating, and more importantly utilising, significant results (something which can be doubted but which requires further study) they represent only a tiny proportion of the industrial population of 6,500 indigenous and 900 overseas companies in Ireland employing more than 3 people. This illustrates the limited potential for small firm and small economy involvement in such technology generation programmes.

In fact a figure which has been quoted for the European Community in general is that less than 1 per cent of firms have been involved to date in all science and technology programmes. This reflects the fact that more than 90 percent of funds are aimed at precompetitive research, while programmes emphasising transfer and diffusion, such as SPRINT, which are aimed at a much wider range of activities and clients, are getting less than 5 per cent of the total resources allocated.

Efforts are being made to raise the level of Irish industrial participation in EC research programmes and also in specially designed bilateral initiatives for the limited number of new technology-based firms (NTBFs) which can potentially involve themselves in such activities. Few Irish firms appear to be participating in research collaborations outside such formal initiatives as the Framework Programme and EUREKA. Interestingly enough, however, and worthy of further research, is the fact that preliminary returns to the most recent biennial survey of technology-related activities (which relates to 1988) indicate that a number (C.30) of foreign subsidiaries consider themselves to be part of research consortia with their parent companies and other subsidiaries. This throws open for debate the whole question of research relationships within multinational companies as they become more globalized. There is also some evidence that while emerging NTBFs in the small countries are being taken over by large foreign firms (as well as being "brought into" through minority equity ownership by companies such as OLIVETTI) they are being allowed to retain their technological and marketing autonomy and have entered into a type of strategic alliance with their purchasers/parents – gaining greatly from the major market contacts of their new "partners". Thus for the smaller countries, ownership may in some ways be becoming a less strategic issue even though many of them are worried by the continuing trend towards mergers and acquisitions internationally. However, the long-term overall effects of this new aspect of internationalization and partnering are still not clear.

Licensing and technology transfer

The first part of this paper made a major issue of seeing international technological cooperation and strategic alliances as encompassing relationships concerned with transfer and diffusion as well as generation. These are often regarded as being activities of lesser worth compared to the phase in which technology is generated but it is here that the real wealth and economic benefits are generated. there is also often the mistake assumption that disembodied technology transfer is a simple straightforward process that almost any company can involve itself in, whether in terms of licensing-in or licensing-out. Our recent surveys of technology transfer patterns in Ireland indicate only a small growth in activity over the 1980s and a relatively small number of indigenous firms involved (C.300). It is generally agreed amongst the "intermediary community", which has grown up in recent years to mediate the transfer process, that the limits are quite finite with only about 200 of those firms currently having the required strategic perspective and complementary assets to be in a position to develop and implement a full licensing strategy. Another 200, some of whom have already been involved in a minor way, need to be consciously assisted over that threshold.

Successful technology transfer usually involves an element of R&D and a number of studies of small firm licensing carried out at the University of Bath in the early 1980s[6] indicated that the most successful firms were those both licensing-in and licensing-out. While the recent research on patent licensing quoted in Section 1 does not find similar evidence in Ireland cases outlined at the beginning of the book indicate that firms involved in European research programmes are also open to, and adept at, other types of cooperative and reciprocal relationships such as licensing. In fact a situation may now be evolving where such firms may be carrying out research for larger firms elsewhere and transferring-in the products they want to produce and market themselves. This illustrates the importance of having a strategic perspective at all stages of technology cooperation and development.

Subcontracting

Just as Japan has become the model for many aspects of industrial and technological policy, so it is with strategic alliances – from research collaboration, through licensing and joint ventures to another form of technological cooperation and internationalization which is now clearly emerging i.e. a new type of subcontracting. What was originally seen as one-way dependency now appears to be changing into a two-way partnership in which even joint process R&D and certainly many other aspects of technological upgrading is going on.

This was also the focus of a new question in the current Irish Business Enterprise Technology Survey where respondents have been asked to indicate if they are involved in subcontracting relationships with companies abroad. To date, more than 100 have indicated that they have relationships which involve taking specifications from an overseas-based contractor. Further studies are necessary to elicit the technology transfer effects of these relationships as the current survey is only aimed

6 For example, Lowe S. and Crawford N.., *Innovation and Technology Transfer for the Growing Firm* Pergamon Press, 1984.

at indicating the population of subcontractors with technology-based relationships.

Section 4: National Responses – Policies and Programmes for Interfirm Cooperation

Policy-making and programme development aimed at encouraging the cooperation process obviously still takes place mainly on the national level. While the regional and international authorities exert an increasing level of influence in all areas of policy and programming national governments still takes most of the decisions which influence firm, including interfirm, behaviour.

It is clear from this section that national governments have made some adjustments to their technological and industrial policies in response to the emergence of the small firm cooperation paradigm. But the important issue here is the need for integrated support programmes and mechanisms as cooperation can only exist when each of the participants has something to offer and something to gain.

The most striking thing perhaps is that, although, for the smaller countries in particular, interfirm cooperation usually means cooperating across national boundaries, relatively few of the programmes that have been designed and implemented fully take into account the broad international context within which they must now operate. Even fewer programmes are specifically aimed at bilateral inter-country or inter-regional cooperation in business and industrial terms, even though scientific exchanges have been taking place for years between higher education and government research institutions under inter-governmental agreements.

A number of the recent Irish initiatives outlined here are exceptions to this rule but these have generally been implemented on a pilot scale and are often, not yet at least, at the centre of national technological and industrial policy. At the same time they point the way forward if they can be moulded into a coherent innovative response to new international conditions for industry and technology.

The section opens with two papers from larger countries – Britain and Germany – which look at the potential and the performance of small firms in national research programmes for particular technologies and in innovation programmes many of which have a collaborative element. They also assess the position of their firms, particularly SMEs, in international programmes like EUREKA. In Britain the approach is to encourage the setting up of research clubs and other networks to encourage small firm participation while in Germany a number of specific small-firms-oriented programmes have been introduced.

Going on then to "smaller" economies, the Spanish (particularly Catalonia) and Portuguese innovation and cooperation approaches are outlined. These show Portugal in the position of getting its research and innovation infrastructure organised in a coherent and consistent way in preparation for a major programme of interfirm and inter-country cooperation. The picture from Spain is one of a "two-speed" economy, with a group of sophisticated firms and universities heavily involved in cooperation within E.C. programmes in particular, and information and technology intermediary organizations being put in place to draw in a wider set of players.

The Irish situation shows a mixture of "old" and "new" policies, with the successful "foreign investment" policies of the past vying with a range of new approaches aimed at indigenous industry and promoting cooperation and partnership as a central focus. While a number of interesting and innovative approaches have been tried these have not yet achieved the critical mass of results needed to change the overall thrust of policy.

None of these countries shows the same awareness of the importance of informal networking as in the Danish situation outlined in Section 3. Here a new technology and industrial policy has been launched which aims to train and employ "network intermediaries". The need for such an approach has been hinted at in number of the other papers in this book but despite current research and the Danish initiatives, the role of such intermediaries has still not been fully defined and understood.

Research Cooperation in the Federal Republic of Germany

A. Goerdeler

In recent years the view has been gaining increasing acceptance in the Federal Republic of Germany that there is a growing need to create the right environment for research and industry in order to stimulate innovation.

Innovation means new and better products, new technologies which increase human productivity, thus opening up a life secure from the material aspect and new opportunities for human beings as well as helping to adjust to structural change in the economy and improving international competitiveness.

A dynamic economic process of innovation that makes long-term growth possible and creates jobs involves basic research on a broad scale, applications-oriented research and technological development as well as the rapid implementation of results within enterprises. Improving in the transfer of both know-how and results between these aspects of the innovation process, through research cooperation, is one of the most important tasks of government technology policy.

The Federal Government encourages research cooperation in different ways:

Improvement of Framework Conditions

The third amendment to the Framework Act for Higher Education 1985 brought about a lasting improvement in the general legal conditions for cooperation between universities and industry. Scientists can now conduct R&D funded by third parties without the approval of their university.

The trend over recent years is encouraging: between 1981 and 1987, the universities received in excess of 330 million DM more in third-party funds from industry and private foundations, amounting to a total of some DM 450 million in 1987. Other research institutions, too, such as the Fraunhofer Institutes, have increasingly conducted research for industry. In 1987, contracts from industry attained a volume of approximately DM 120 million (i.e. more than twice as much as in 1982). About two thirds of the 600 customers were SMEs.

Collaborative Research

The promotion of collaborative research, which aims in particular to bring about increased cooperation between research institutions and industry, has been expanded since 1984. Now, some 60% of the funds provided by the BMFT (Ministry for Reasearch and Technology) for the direct promotion of market-oriented technologies is accounted for by collaborative research projects (1988: 1 billion DM). The expansion of collaborative research has taken place predominantly in areas of key technologies (production engineering, information technology, materials research etc.). The experience so far of those involved in the projects shows that joint integrated work, although time-consuming, opens up the way to new contracts, facilitates the transfer of technology and personnel, brings about a concentration of research potential, produces critical mass and triggers synergistic effects.

EUREKA

The EUREKA initiative, too, is based on the idea of collaborative research. EUREKA has created a framework for European cooperation between companies and research institutions from different countries. According to the bottom-up approach adopted, the project partners themselves determine, beyond the bounds of national frontiers, how, on what and with whom they wish to cooperate. These activities are accompanied, wherever possible and necessary, by administrative and political support measures – for instance, the elimination of trade restrictions or the laying down ofjoint European standards. Companies participating in EUREKA are supported in the Federal Republic of Germany by the usual programme-related project promotion.

Cooperative Industrial Research

Intensive cooperation between research institutions and industry takes place within the framework of cooperation in industrial research, which has a long tradition in the Federal Republic of Germany. The Confederation of Industrial Research Associations (AIF), founded in 1954, now comprises 93 associations from 32 industrial branches and 63 research institutes. Cooperative industrial research involves the performance of research projects by the institutes of the AIF and also by research institutions at the universities, at the Max Planck Society etc. These projects reflect the common interests of the members of a particular sector-specific association and are intended to induce the practical application of basic scientific and technological knowledge.

Contract Research

In contrast to the basic questions which can be solved by cooperative research at the industrial level, external contract research as a tool will help to overcome individual R&D bottlenecks in SMEs. If, due to lack of experience or personnel, a company cannot solve specific R&D problems itself, very often R&D contracts with research institutes or engineering bureaux are a useful approach. In order to overcome existing barriers to the placing of R&D contracts with external partners, BMFT is providing special support amounting to 40% of the expenditure on R&D contracts for companies with a turnover of less than 50 million DM, up to a maximum amount of 150,000 DM per firm.

Placement of R&D Personnel in Research Institutes

In order to stimulate and increase "technology transfer by heads and brains", which is one of the most important methods of technology transfer, BMFT started a special programme in 1984 to enable young industrial scientists and engineers to work for a certain time at universities or government institutes in the areas of interest to their company and to return to industry having acquired new knowledge, skills and experience. The Federal Government supports these scientists and engineers by financing part of their salaries for a period of up to 3 years. An encouraging aspect is that SMEs, in particular, are availing of the scheme.

Young Technology-based Start-up Companies

A new pilot scheme has been launched by the BMFT this year to stimulate new forms of cooperation between founders of new firms and financiers. The new scheme comprises two variants.

- One provides for renewed financing of investors who involve themselves in the foundation of new technology-based firms. The risk of non-payment on the part of the investors is met chiefly by a public banking institution.
- Under the second (the co-investor model), a different public banking institution makes capital directly available to the founders of new technology-based firms, provided that other investors commit themselves to at least the same amount. In this case, the risk of non-payment on the part of the investors is also met chiefly by the public banking institution. BMFT promotion under the new pilot scheme can be seen in that the interest to be paid and the risk of non-payment on the part of the two public banking institutions involved are shouldered to a major extent by BMFT.

Establishment of Technology Transfer Bureaux and Innovation Advisory Services

The Chambers of Commerce and other industrial associations or foundations have set up a network of technology transfer advisory boards and networks of information banks during the last decade. Initially financed by BMFT during the pilot phase, they are now more or less self-financing through selling their services. Their aims are to provide information on the funding of SMEs with regard to new technologies (e.g. microelectronics), the organization of contacts between SMEs and experienced R&D institutions, information brokerage for specialized literature and databank systems and advice on applications for government R&D support.

Information and Guidance in New Areas of Technology

In areas of new technology, a start has recently been made on promoting "demonstration centres" on a temporary, pilot basis. Such centres inform potential users, without reference to specific producers, about the possibilities for applications and the opportunities for implementing new technologies, thus helping to reduce the initial risks for SMEs when embarking upon such new technologies.

Furthermore, they promote the rapid development of a market for advisory services and training because they can be accessed also by private advisory services and vocational education organizations. Demonstration centres have been set up in the areas of production engineering, information technology and new materials.

To sum up, there is now a widespread network of R&D institutions and industrial organizations in the Federal Republic of Germany which reinforceg the links between companies themselves and between companies and science. The Federal Government will endeavour to further such cooperation in R&D. The urgent need for government commitment is underlined by the growing complexity of the innovation process and by the interface problems between the acquisition and commercialization of new technological knowledge. A further factor is the increasing interlinking of different areas of technology (e.g. mechanical engineering, electronics and sensors), which is beyond the scope of individual companies, particularly SMEs, and therefore necessitates cooperation.

Small Firms – the Impact of European and UK Policies

B. R. Arthur

Introduction

Small and medium-sized enterprises (SMEs) form an important sector of any developed country's industrial structure. In the UK, SMEs employ the majority of people working in the industrial sector and make an important contribution to innovation, new products and market growth. The UK has encouraged the growth of new companies and entrepreneurs by reducing barriers to the formation and growth of companies, reducing bureaucracy and establishing an open market economy. The basic assumption behind the policy is that the way to wealth creation lies through business and individuals working efficiently in open competitive markets with economic decisions being best taken by those competing in that marketplace.

The responsibility of government is to create the right climate so that markets work better. For the Department of Trade and Industry (DTI) in particular to encourage the "enterprise culture". DTI's Innovation Policy is focused primarily on circumstances where research is necessary before commercial applications can be developed and thus intervention in the market is justified. As a result there is now much greater emphasis on longer term pre-competitive collaborative research. Collaboration is encouraged in order that the benefits of government intervention and grants should be spread widely throughout the relevant sector. This is why the DTI also focuses on SMEs to ensure that the latest technologies and best practices reach all parts of the industrial market.

The DTI Innovation Policy has four specific objectives. These are:

- to encourage industry to increase its funding of research and development and to apply new technologies more effectively;
- to encourage the translation of inventive capability and best practice into commercial application by promoting greater and better technology transfer;
- to encourage industry to make most effective use of its own and other sources of information through collaborative research and development, both nationally and internationally;
- to encourage innovation by small companies especially in advanced technologies and in regions of higher unemployment.

DTI works with other government departments in the execution of its innovation policy, most notably the Department of Education and Science and with the science base in the Higher Education Institutes (HEIs to promote the fuller exploitation of publicly funded research and development.

This policy was introduced early in 1988 by the then Secretary of State Lord Young, in the White Paper on the Enterprise Initiative. It emphasizes long term collaborative research between companies, collaboration between higher education institutes and companies and the facilitating of technology transfer. It reflects a move away from near market, general schemes providing grant assistance to individual companies. The policy recognizes the special importance and the role of SMEs, particularly in collaborative R&D and technology transfer.

For the purpose of this paper I will split the DTI support for SMEs between international and national schemes and then briefly mention other government schemes which encourage collaborative ventures involving SMEs.

International Programmes

There are a wide range of Community and international programmes in which SMEs can participate. A detailed description of each programme is beyond the scope of this paper and I will therefore concentrate on those programmes of most importance to SMEs in the high technology sectors where Community or government intervention on an international or a national scale can be justified.

EUREKA

EUREKA is an international programme sponsored by national governments which promotes collaboration between companies in the EUREKA member states which include EC, EFTA and others. The EUREKA programme has now supported some 297 international collaborative projects, total cost £4b Ecu. The UK is represented in 86 of these projects (which total £1.1b Ecu) by 160 companies of which 25 are SMEs.

UK SMEs have not encountered any inherent obstacles in the EUREKA mechanism over and above those normally associated with small companies taking part in international projects. Most UK SMEs are involved in user/supplier or user/producer vertical collaborations. There are a small number of international SME to SME EUREKA projects.

A study some two years ago indicated that approximately two thirds of all EUREKA projects included SMEs and that between one quarter and one third of the companies involved in EUREKA had under 500 employees.

Unlike other international collaborative and UK national collaborative programmes EUREKA does allow collaboration on closer to the market projects. This is very helpful to smaller companies and to some extent offsets the natural difficulties for small companies collaborating across national boundaries.

ESPRIT

As delegates will know ESPRIT I and II are the EC Framework R&D Programmes supporting collaborative research in Information Technology. In ESPRIT I approximately 14% of funding went to SMEs, although 60 to 70% of the projects included a smaller company. The figures for ESPRIT II will probably not show great change despite the increased efforts that the Commission has made to encourage the participation of European SMEs.

A survey of SMEs which participated in ESPRIT I indicated that 25% of SME's were well pleased with the outcome of their projects and a further 50% felt they got an adequate return for their investment in the collaborative R&D work. Thus although SME's have not always found it easy to participate in ESPRIT – most companies comments cite the long timescales and the amount of effort necessary to finalize projects – it appears that of those small companies that have taken part three quarters have considered their projects worthwhile.

Although the Commission continues to make strenuous efforts to increase the participation of SMEs in its R&D programmes it is, in my judgement, unlikely that the ratios will change significantly. EC R&D programmes are formulated in order to undertake only that collaborative R&D which is best done by Community-wide action. The Framework Programmes are not intended to duplicate national programmes or areas of R&D which can be adequately carried out within individual member states.

BRITE AND EURAM

The BRITE, and latterly combined BRITE/EURAM programme, like other Community R&D programmes has made no special concessions for the participation of SMEs. Projects proposals are assessed on their technical/scientific excellence, appropriateness of research and collaboration, and exploitation route.

There has been strong participation from smaller UKcompanies, especially so in the later stages of the programmes. For example in BRITE the first "call for proposals" resulted in SMEs forming 30% of the companies participating – although only 15% of UK companies were SMEs. In the second call 41% of companies participating were SMEs overall and of the UK companies participating 40% were smaller companies. Industrial companies represented 60% of all participants in the second round.

As with other international programmes UK SME' have not always found it easy to participate in the programme – it requiring a high degree of dedication and tenacity to complete the project negotiations. Those companies that complete the course generally believe their participation to have been of net benefit to their organizations.

UK National Programmes

The LINK Scheme

The LINK scheme is an inter-departmental initiative which seeks to build bridges between UK industry and the science base so that scientific research and knowledge can be transferred to industry and subsequently exploited in the marketplace. LINK involves all UK government departments which fund research, most notably the DTI and the Department of Education and Science, DES. The LINK scheme encourages companies to undertake collaborative research in a range of selected strategic topics across a wide range of disciplines. The Scheme is organized through a series of programmes each chosen as an important emerging area of knowledge. It is usually located in the HEIs and the research is at a precompetitive stage. Currently 22 programmes have been announced (see Appendix 1 to this paper for list) covering a wide range of topics in electronics and communications, food and bio-sciences, materials and chemicals, engineering, and measurement and sensing.

Companies participate in the programmes by taking part in collaborative projects normally involving at least one HEI and two companies. There are special relaxations for SMEs in order to expedite their involvement in the programme which allows "one on one" collaborations between auniversity and an SME.

As yet it is early days in LINK: of the over 500 projects at various stages of preparation 41 are active so far. Of the 73 companies taking part in these active projects 16 are SMEs (22%).

SME participation in LINK programmes is mixed since there is a very wide range of scientific topics included across the scheme. Some programmes, for example the Industrial Measurement Scheme, are aimed at an industrial sector heavily represented by SMEs and the majority of projects therefore include at least one smaller company. In other programmes like Molecular Electronics, which is still a long way from any commercial realization, it has been more difficult to encourage smaller companies to participate.

The Science and Engineering Research Council, SERC, and DTI recently commissioned a study undertaken by Michael Shortland Associates to report on the potential involvement of SMEs in the LINK Molecular Electronics programme. This showed that despite extensive publicity and awareness activities many SMEs in the target industrial sectors were unaware of the Molecular Electronic LINK programme and most only vaguely aware of LINK itself. Many of those target SMEs contacted in the survey expressed a positive interest and many were keen to join into the programme immediately. Whether those companies will actually participate is another question. The report therefore suggests that despite extensive efforts to publicize collaborative R&D programmes the message is often still not getting through to the smaller companies in the target industrial sector. It suggests that a much more interdictive "search and follow up" activity is required to identify and interest smaller companies in these programmes.

Advanced Technology Programmes

The DTI Advanced Technology Programmes are a part of the DTI Research and Technology Initiative. They aim to encourage pre-competitive research in new technologies from industrial companies working in collaboration. The objective is to build up UK industrial capabilities and accelerate the applications of these technologies in new and existing markets. The AT Programme consists of 8 self contained programmes of research, development and technology transfer in new important technologies.

Programmes consist of a number of related projects which must have at least three collaborators, two of which must be able to exploit the outcome of the research. Funding is negotiated at up to 50% of total project costs and is usually tapered, reducing over the life of the project.

The programmes are described in Appendix 2 to this paper.

SMEs are represented in all AT programmes. Projects involving large companies are encouraged to include SMEs as part of their collaboration. Differential grant rates between smaller and larger companies may also be used to enable SMEs to participate in projects.

General Industrial Collaborative Programme

The General Industrial Collaborative Programme, GICP is the third of the DTI national collaborative initiatives. This supports research projects (and small programmes) which do not fit into the major strategic programmes of LINK and the Advanced Technologies Programme. The programme also supports many "Club" type projects where the research is carried out by a Research and Technology Organization, RTO, on behalf of a number of companies. The results of the programme are then freely available to all of the participating companies. This form of collaboration is especially helpful to SMEs as it allows them to access advanced science and technology without requiring research facilities and personnel of their own.

In addition the programme supports a number of projects involving SME's in both vertical and horizontal collaborations in a wide range of industrial sectors. This programme is also aimed to encouraging companies in more traditional "Low-Tech" industries to invest in research and echnology. The UK AIRTO (Association of Independent Research and Technology Organizations) recently reported that over 80% of its members were companies employing less than 500 people and that 67% of its members were companies employing less than 200 persons. The RTOs therefore represent a powerful mechanism for involving SMEs in collaborative and cooperative ventures.

Other Government Department Schemes

In this section I shall mention, only briefly, three other schemes which encourage cooperation involving SMEs of which I am aware.

Teaching Company Scheme

This is a joint SERC/DTI initiative which encourages academics to work in industrial companies, particularly SMEs, for a limited time. The objectives for the programme are to expose academics to the industrial, manufacturing environment and to transfer relevant university know-how into the company.

Cooperative Grant Awards

There are two schemes run by the SERC and the Ministry of Defence which encourage collaboration between universities and firms including SMEs, for collaborative research projects. Both schemes work on a "one on one" mode, that is one company collaborating with a university.

Conclusions

In the UK a wide range of manufacturing SMEs can participate in gvernment grant-aided collaborative research projects. The largest proportion of SMEs participating in UK programmes is in IT, specifically software-related specializations. This is perhaps not surprising since this area of technology is particularly appropriate to smaller companies. Those SMEs which have successfully completed the long negotiating stages leading up to a collaborative project have generally found a net benefit to their company in new technology captured, new business and market contacts made and knowledge gained.

Despite the best efforts of programme designers and policy-makers SMEs still appear to be unaware of the existence of government-funded collaborative research programmes aimed at their industrial sector. DTI is working together with other agencies like AIRTO to address this problem.

UK SMEs which have sought to participate in European programmes have been relatively successful in their applications and in the outcome of their projects. Nevertheless international collaboration – whether through Community programmes or through EUREKA – remains a severe test of stamina for any smaller company.

Appendix 1

Link: Current Programmes (by field)

Electronics and Communications	Prog. No.	Date Ann.
Molecular Electronics	1	3/2/88
Advanced Semiconductor Materials	2	3/2/88
Personal Communications	7	13/6/88
Optoelectronic Systems	12	26/4/88
Power Electronic Devices & Derived Systems	18	27/7/89
Food And Bio-sciences		
Eukaryotic Genetic Engineering	10	3/2/88
Biotransformations	6	15/4/88
Selective Drug Delivery and Targeting	8	14/6/88
Food Processing Sciences	10	7/9/88
Protein Engineering	15	12/6/89
Control of Plant Metabolism	19	19/9/89
Materials and Chemicals		
Structural Composites	14	16/5/89
New Catalysts and Catalytic Processes	17	19/6/89
Engineering		
Nanotechnology	5	3/2/88
Construction, Maintenance & Refurbishment	9	14/6/88
Ventilation, Air Conditioning & Refridgeration	11	17/4/89
Design of High Speed Machinery	13	10/5/89
Biochemical Engineering	16	15/6/89
Transport Infrastructure and Operations	22	?
Measurement and Sensing		
Industrial Measurement Systems	3	3/2/88
Technology for Analytical & Physical Measurement	21	27/9/89
Molecular Sensors	20	27/9/89

Appendix 2

1. **Wealth from the Oceans** – sea bed mining, exploration and exploitation (not fish farming).
2. **High Temperature Superconductivity** – Research on HTS materials, applications and theory.
3. **Gallium Arsenide** – Gallium Arsenide chip technology and applications research.
4. **National Electronics Research Initiatives** – Four projects in advanced microelectronics research in Silicon Superhybrids, Pattern Recognition and Vision Systems, Silicon Architectures CAD software research, Advanced Lithography for semiconductors. Work carried out by collaborators at host laboratory.
5. **Information Engineering AT** – IT and related technologies in system engineering, systems architectures and CAD and Hardware.
6. **Advanced Robotics** – Collaborative research in all aspects of advanced robotics and robotic vehicles.
7. **Computer Aided Engineering** – Applications of vision and artificial intelligence to manufacturing technology. Modelling and simulation codes of best practice etc.
8. **Systems Technology and Integration** – Software systems for manufacturing MAP/TOP extensions and developments of computer communications protocols for SME standards and integration for software into systems.

Portuguese SMEs in the European Context – Prospects for Innovation and Cooperation

I. Matalonga, R. Guimaraes, M. de Abreu, A.P. Coelho

Introduction

The changes in and evolution of the international market trends in the 1970s, induced a strong consciousness in the industrialized world on the need for technological innovation in production process and product definition.

This is a rather demanding challenge, because the very nature of the organization, in terms of management and production, has to go through a substantial resettlement. So it is not hard to understand why SME's have been facing great difficulties in keeping pace with the adjustment process. For instance, compared with what has happened in other geographical areas, one may find, a restrictive economic context and a strong conservative nature in Europe in what concerns entrepreneurial behaviour. These act as negative elements in Europe's industrialization process.

European SME's have been aware of their different problems and are using several measures and approaches, which sometimes include suitable cooperation and interaction between firms, to promote their technological development. Latterly, the same process has happened at the Community level but has still not proved to be successful.

An essential issue that should concern all major European decisions, at all levels, is to get an effective collaboration among European SME's in different areas, an action that involves their business and technological development.

The Portuguese Industrial Structure

The Portuguese industrial structure in the 1970's, and even today, conforms with the structure of other small economies operating in the European context.

In Portugal, manufacturing industry is still largely dominated by SME's which for the most part are on a very small scale. On the other hand, industrial production is based upon light manufacturing goods which depend on intensive labour technologies involving low paid and semi-qualified workers.

Within this framework – reinforced by a strong sense of individuality, which is a common feature of "small and medium" entrepreneurs in small economies – it was not easy to lead the firms into a cooperation venture, in order to foster innovation and upgrading.

However, taking advantage of a certain degree of flexibility which is also an SME feature, innovation could be stimulated, provided that there was the scientific and technological basis from which SMEs could move to an effective development process.

For a long time this type of process has not been fully attained in Portugal since the necessary scientific and technological infrastructures have not been put in place. Only in the decade starting in 1976 was a strong effort carried out to build up an effective system in this area and also to establish the suitable bodies and agencies which could deal with applied research and technological development issues.

Since then, one can observe the beginnings of a successful performance. Even today, a balanced survey of the Portuguese scientific and technological systems and capabilities would conclude the following:

- a lack of human and financial resources allocated to R&D activities (in 1986 R&D investment was only 0.46% GDP);
- insufficiently built-up infrastructures, sometimes with outdated equipment and laboratory facilities;
- industry strongly tied to external markets which involve the acquisition of technology and capital goods;
- a low rate of enterprises investing in R&D activities;
- at all levels, deficiencies in scientific and technological training.

This situation is set to change very quickly and there is currently a strong will on the part of the universities and public laboratories to cooperate with SMEs and to respond adequately to their needs.

Antecedents

In the early 70's while primary production remained stable, manufacturing industry registered a sudden increase of 10% on average in the period 1970 - 1973. If this boom can be seen as the result of a significant number of investments, due to the action of internal economic groups and foreign investment, it can also be a consequence of light export industries turning towards EFTA markets. In 1973 SMEs in textiles, food industry, leather goods and footwear, cork and wood accounted for more than 40% of manufacturing industry output.

In the second half of the 70's the growth of exporting SMEs was very important to the Portuguese economy. In fact in mid-decade, strong internal changes, reinforced by the international energy crisis led to a whole new atmosphere of depression.

In these circumstances the most important basic industrial companies (some of them public) proved to be unable to adjust to the new national and international situation. On the contrary SMEs in the light exporting industries, although facing technological and management difficulties, have often adapted themselves to the changing pattern of demand and have also acquired a privileged status in the context of the Portuguese economy.

In 1982, out of a total of 13,536 industrial enterprises, 60% had less than 20 workers per enterprise, and 80% had less than 50 workers. So it is not strange that the Portuguese Government has paid prime attention to SME problems and expectations in the definition of its industrial and technological policies.

Changing Environment (1975-1985)

In 1975 the Ministry of Industry set up a new institution IAPMEI (SME Support Institute) oriented mainly towards supporting and upgrading of SMEs.

IAPMEI's activities are directed at four areas:

- management upgrading through training and direct company assessment;
- intervention in the investment and re-equipment programmes of firms through acting as a channel to credit and banking services;
- supporting marketing studies and other SME-orientated marketing initiatives;
- promoting cooperation between firms and contributing to actions aiming at the reorganization of traditional light industrial sectors.

In 1976 Portugal already trading almost exclusively with European countries, proposed to enter the EC. The Government, concerned about the technological gaps, between Portugal and other European countries, decided to widen the scope of its technological policies by creating a specific body – LNETI, created in 1979 – which would be able to deal with the technological needs of Portuguese industry.

Since then LNETI has concentrated its activities on the technological development of Portuguese industry, stimulating technological innovation through the development of new processes, products and prototypes on the basis of contracts with firms.

The principle activities of LNETI (National Laboratory for Engineering and Industrial Technology) are:

- to design and implement Research, Development and Demonstration (R, D&D) programmes and projects for the modernization of Portuguese industry, related to strategic areas defined by the Portuguese Government;
- to foster new industries and cooperate in their implementation, mainly those concerned with advanced technologies;
- to develop or adapt new technologies and introduce them to enterprises with the aim of increasing competitiveness and profitability.

The Technological Development Plan – the Role of LNETI

In the early 1980s LNETI's experience led the Portuguese Government to set up a Technological Development Plan for Manufacturing Industry. This plan (approved in 1983) defined the following programmes of action:

a. Reinforcement of Technological and Scientific Capabilities;

b. Promotion of Technological Improvement and Industrial Innovation;

c. Sectoral Programmes for Technological Development;

d. New Technologies: Acquisition, Demonstration and Training;

e. Creation of High-Tech Intensive Enterprises;

f. Establishment of an Intermediary Financing Agency (FAIT);

g. Several Support actions (e.g. – technical training, specific programmes).

To achieve the objectives of these programmes, as clearly defined in the Plan, a combined effort from both industry and government was necessary. The primary objectives were:

- the modernization of existing industrial sectors;
- the introduction of new enterprises on a technological basis;
- the reinforcement of the scientific and technological structure which includes the training of human resources in the new technological sectors.

This plan has been worked out in consultation with the leadership of LNETI which has also been taking an active part in its implementation through the following:

- technological development of traditional industrial sectors, in particular through the establishment of technological centres. These are connected with the modernization and improvement of specific industrial sectors. The active participation of LNETI, IAPMEI and several Industrial Associations in these centres is also important. The following ones are actually operating: CATIM – Technology Support Centre to the Metalomechanics Industry; CTCV – Glass and Ceramics Technological Centre and the CTC – Footwear Technology Centre;
- acquisition and demonstration of new technologies namely through the creation of enterprises for research and development. Currently, there are two companies of this type, in the fields of Information Technologies and Biotechnology;
- the establishment of financial incentives to support industrial innovation, namely: Industrial Development Contracts; – CDI's, which involve cooperation between universities, public laboratories and enterprises.

Within the spirit of the Technological Development Plan, some technology-based companies have been founded, using "incubators" and risk capital firms. These are instruments of areal strategy to infuse dynamism into the evolution of the Portuguese industrial sector at the SME level, especially those involved in Advanced Technology. (See Appendix 1 to this paper).

These policies helped to create a propitious atmosphere for innovation in SMEs. At the time when Portugal was about to become a member ot the European Common Market

The Specific Programme for the Development of the Portuguese Industry (PEDIP) – the Involvement of LNETI

By joining the EC it was possible for the Portuguese Government, with the support of the Commission, to design and implement a new whole set of programmes and actions which were integrated into the "Specific Programme for the Development of Portuguese Industry – PEDIP" focusing particularly on the innovation process and on general technological upgrading. In many cases, detailed and precise attention is given to thesmall, specific investments made by SMEs.

Within the clear objective of influencing a wide range of sectors, four main priority areas were set up in order to achieve the above mentioned objectives.

Two of them are aimed at reinforcing existing infrastructures and increasing the level of advanced training actions; the others are aimed directly at financing and productivity problems in enterprises.

Within this context seven programmes have been set up, namely:
Programme 1 – Technological and Basic Infrastructures
Programme 2 – Advanced Technological Training
Programme 3 – Incentives to Productive Investment
Programme 4 – Financial Engineering
Programme 5 – Productivity Missions
Programme 6 – Quality Missions and Industrial Design
Programme 7 – Diffusion, Implementation and Control
(See Appendix 2 to this paper)

LNETI is deeply involved in the implementation and evaluation of different PEDIP programmes, mainly in the areas of :

- Technological Infrastructures;
- Advanced Training on New Technologies;
- Information Technologies.

LNETI itself is about to develop a set of technological and demonstration units which are linked to innovation processes in light industry i.e. wood and cork, food industries, cad/cam, biotechnology, ceramic materials and environmental technology.

A contribution of considerable importance to the development of new SMEs, especially in the technological area, will be the future performance and progressive building up of the new Lumiar Technopole which will involve industrial SMEs and research centres as well as a certain number of "incubators". This Technopole will be built next to the LNETIs Campus, in Lumiar (Lisbon).

EC – Perspectives for Portugal and its SMEs

Joining the EC has created new perspectives for industrial firms regarding international cooperation.

Nevertheless, due to the lack of tradition in research and experimental development and the dominance of traditional sectors of manufacturing industry the opportunities for the Portugese participation in Community programmes of R&D remain at a low level.

Here one must not forget that SMEs and industry in general have experienced some national social and economic instability since the mid 1970s.

This situation has not encouraged a faster pace of industrial adjustment and reorganization. Recently, it has been possible for industry to proceed in the modernization of its production processes and management methods. Thus, it is necessary to benefit from the current strong support to investment in R&D, as an essential factor in order to renew and retain competitiveness in the enlarged Internal Market.

More intensive cooperation between universities, public laboratories and SMEs is vital to stimulate these enterprises into participation in Community R&D programmes. The factors required by these programmes to accomplish their objectives are not available where SMEs are operating in isolation from the other actors. Thus, the setting up of European networks for scientific and technological collaboration should be strongly supported and encouraged.

As we have already stated, the Commission has supported the implementation of PEDIP the operational programmes of which involve specific lines of assistance not only to innovation and the upgrading of SMEs, but also to creating effective partnerships in the Community programmes of an industrial and technological nature.

As far as possible it will also be important to start fighting a perverse effect of our entrance into the EC: the progressive connection of Portuguese universities with non-Portuguese firms within the EC's R&D programmes.

In Portugal one can observe a major characteristic of small countries: R&D activities tend to adopt priorities pre-established by the international scientific community, led by the larger countries.

Consequently, the work of the national scientific community is commercialised by the enterprises of countries possessing wider capabilities. While not putting an end to this type of connection, which gives us prestige and pride, one must also insert the Portuguese enterprises into the circuit.

Recent enquiries under the auspices of the Ministry of Industry and Energy, reached the conclusion that little is known by Portuguese enterprises of Community R&D programmes with two exceptions: EUREKA and the "demonstration projects on non-nuclear energy".

The Energy and Raw Materials programmes represent 75% of total Community assistance. The major restrictions arise, as the enquiry reveals, from the lack of R&D structures (human and material) in the companies and also from the fact that they do not have any strategic plans for development. It is necessary to fight these situations through:

- a specific programme to support the participation of Portuguese firms in Community programmes that will include dissemenation of information and support in making applications
- financial incentives for promoting cooperation between enterprises/research centres and the creation of R&D laboratories in firms through some system of promoting technology development.
- a Specific Programme to put researchers into industrial enterprises.

Appendix 1

Technological Development Plan (1983 - 1993)

A. REINFORCEMENT OF TECHNOLOGICAL AND SCIENTIFIC CAPABILITIES
1. Studies on Technological Policy
2. Setting up of Capabilities in General Technologies
3. Training of Technicians Responsible for Research Projects
4. Connection Structures Between Industry and Research Institutions
5. Creation of an Agency for the Development of Advanced Technologies

B. PROMOTION TO THE TECHNOLOGICAL IMPROVEMENT AND INNOVATION
1. National Design Centre
2. Industrial Design
3. Innovation Centre
4. Information and Industrial Extension Nets

C. SECTORIAL PROGRAMMES FOR TECHNOLOGICAL DEVELOPMENT
1. Technological Centres
2. Consultancy and Engineering Sectorial Enterprises
3. Quality Control R&D, and Training in Enterprises
4. National Program of Equipment Development
5. Modernization & Reorganization of Traditional Branches

D. NEW TECHNOLOGIES: ACQUISITION, DEMONSTRATION & TRAINING
1. Opportunity Studies
2. Combined Actions
3. Demonstration Programmes
4. Training of Scientists and Engineers

E. CREATION OF ENTERPRISES OF TECHNOLOGICAL BASIS
1. Creation of Enterprises of Technological Basis
2. Establishment of Technopoles

F. FINICIAL INTERMEDIARY FOR INNOVATION
1. Creation and Capitalization of the Fund for Support to Technological Innovation

Appendix 2

Programme Characterization

Subprograme 1.1	**Base Infrastructures**
Intervention Area	Infrastructures of Industry's support
Goals	- Support the rationalization of distribution circuits of materials and and industrial production - Promote the industrila accessibility regarding transportation systems
Programme Measures	- Highway infrastructures - Railway infrastructures - Port infrastructures - Support infrastructures to the industrial activity
Direct Beneficiaries	Industrial enterprises
LNETI's Role	LNETI as the promoter of projects

Subprograme 1.2	**Technological Infrastructures**
Intervention Area	Infrastructures of Industry's support
Goals	- Develop the technical and technological capacity regarding the entrepreneurial capacity increase - Foment interinstitutional R, D&D aiming at innovation's introduction in the firms - Intensify interaction between University/Research Centres/ Enterprises - Guarantee conditions to create innovative enterprises
Programme Measures	--Technological support to industrial sectors - New technologies development - New technologies transference - Support to the creation and installation of new technologies entities
Direct Beneficiaries	Industrial enterprises with innovation ptojects in new technologies and new systems or processes
LNETI's Role	LNETI as the promoter of projects LNETI as the valuator of the scientific interest and economic feasibility of the projects

Programme 2	**Advanced Technical Formation**
Intervention Area	Professional Formation
Goals	Should be complementary to the other PEDIP programmes in the Portuguese industry development, through the human resources valorization
Programme Measures	-Several formation activities in the areas of Management, New Technologies, Research, etc.
Direct Beneficiaries	Enterprises or industrial organizations
LNETI's Role	LNETI as the promoter of projects LNETI as the valuator of projects LNETI responsible for the implementation, accompaniment and control of projects

Programme 3	**Productive Investment Incentive**
Intervention Area	Productive investment financing
Goals	Stimulate, through subsidies, the industrial enterprises investments geared to modernaization, innovation and technological development, product quality, energetic restructuirsation and environmental protection
Programme Measures	- PEDIP incentives system (SINFEDIP) - Financing specific operations of the incentives system to the rational use of energy (SIURE) - Support to the restructuring or modernization of industrial sectors - Support to specific sectors
Direct Beneficiaries	Industrial Enterprises
LNETI's Role	In SINPEDIP, LNETI verifies the importance of the project to the industrial development strategy, the access conditions and the purpose the incentive subsidies

Programme 4	**Finicial Engeneering**
Intervention Area	Productive investment financing
Goals	Creation of reciprocal guarantee mechanisms to small and medium enterprises, in which the risk is suppported
Programme Measures	-Grants for the development of : - Risk Capital Associations - Enterprise Foment Assoctiations
Direct Beneficiaries	Industrial Enterprises (namely small and medium enterprises)
LNETI's Role	

Programme 5	**Productivity Missions**
Intervention Area	Missions of Productivity and Industrial Quality
Goals	This programme is complementary to other measures intended to reinforce industrial enterprises' productivity and acting upon dynamic factors of productivity, such as: production management and organizaion, distribution, energy, quality, etc..
Programme Measures	--Demonstration, divulgation and promotion (of ideas and actions promoting productivity increase) - Support to the productivity Management
Direct Beneficiaries	Enterprises (Services and Industry); Non Profitable Institutions; Others
LNETI's Role	

Programme 6	**Industrial Design and Quality Missions**
Intervention Area	Missions of Productivity and Industrial Quality
Goals	- Reinforcement of the National System of Quality Management - Creation of a national awareness either in producers of in consumers of Quality and industrial design - Development of a capacity for Quality Management and Industrial Design inside the industrial enterprises..
Programme Measures	- Quality incentives in the context of the Quality Management National Sytstem structures' support - Other measures concerning quality support, such as: motivation campaigns, increase of the normative activity, etc.
Direct Beneficiaries	Industrial Enterprises; Metrological Laboratories and Non-profitable Entities; Trade Unions; Professional Associations and Associations of Enterprises, etc
LNETI's Role	

Programme 7	**Divulgation, Implementation and Regulation**
Intervention Area	
Goals	Activate the divulgation, implementation and control mechanisms of the several PEDIP programmes
Programme Measures.	
Direct Beneficiaries	All the beneficiaries of the other PEDIP programmes
LNETI's Role	

Budget in MECUs

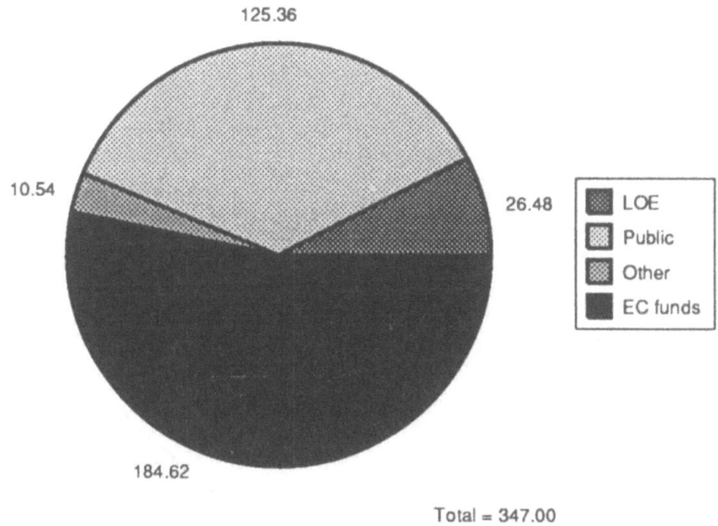

Chart 1: Subprogramme 1.1

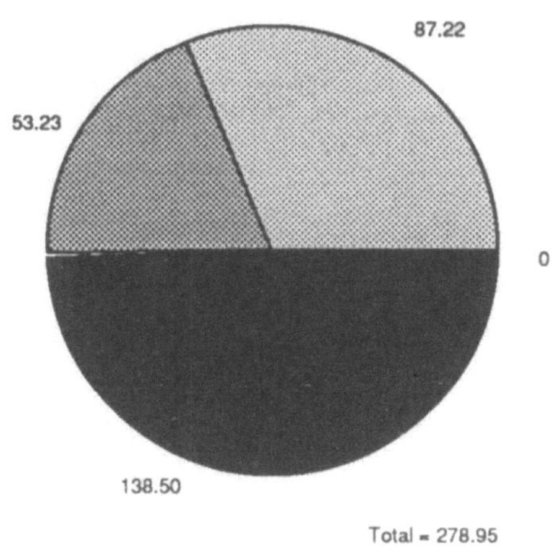

Chart 2: Subprogramme 1.2

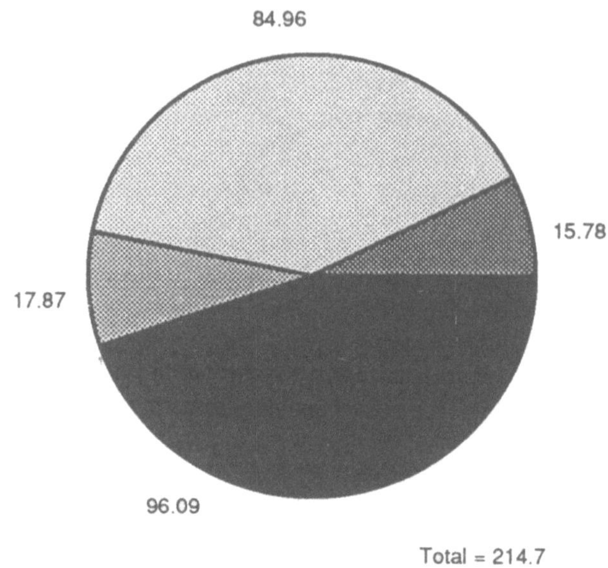

Chart 3: Programme 2

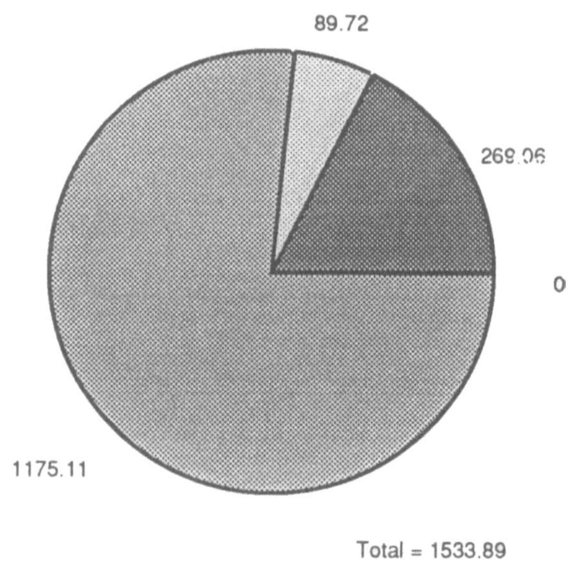

Chart 4: Program 3

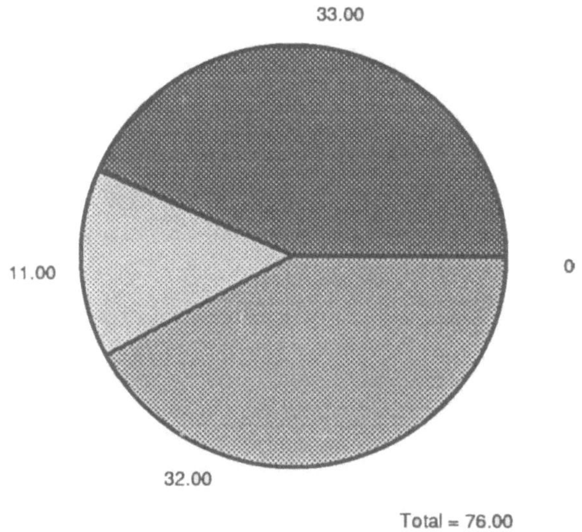

Chart 5: Programme 4

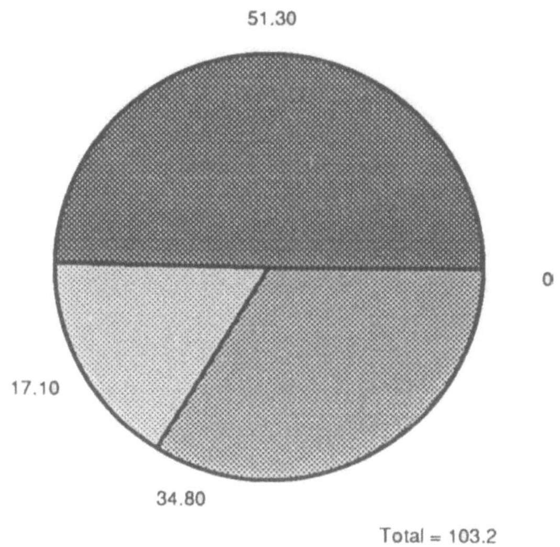

Chart 6: Programme 5

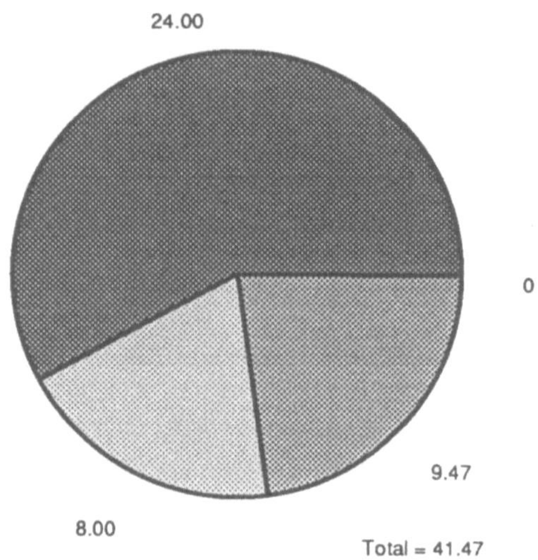

Chart 7: Programme 6

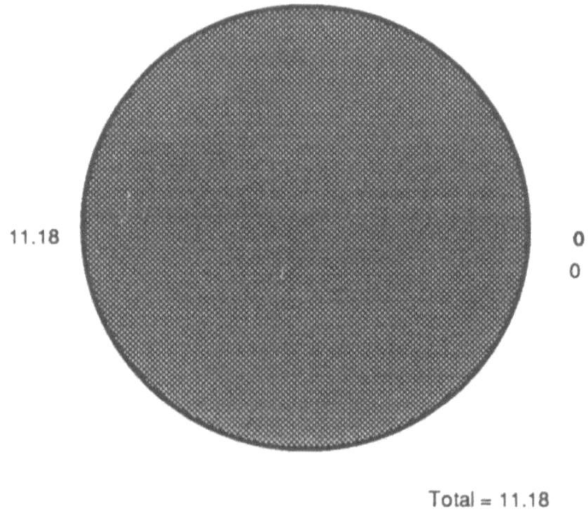

Chart 8: Programme 7

Cooperation in Spain – the Role of Public Institutions

J. Valls Passola

Introduction

Since 1985 the Spanish economy has experienced an obvious recovery. Thus, in the recent years its growth-rates have exceeded the average in Western countries. Moreover, waves of foreign investment have been continuous. Our entry into the EEC in 1986 has given us the reputation of being a dynamic country, even though, globally, the infrastructures are not at the level they should be and the regional imbalance is, as everybody knows, considerable.

Because of regional diversity and the differing economic situations in Spain, there are also diverse industrial issues resulting from very different business and entrepreneurial traditions. We consider this preliminary observation necessary because most of this exposition applies specially to Catalonia (6 million inhabitants, 17% of the Spanish GNP). The innovative activities of Catalan firms have been analyzed on several occasions in the Business and Economics Department at the Politechnical University of Catalonia in recent years.

The Perception of Cooperation

Focusing specifically on technological cooperation we can see that in Spain, as elsewhere, companies have slowly introduced cooperation as a strategic necessity in the context of the growing internationalization of the economy. In the case of Spain, three kinds of cooperation can be distinguished:

First, for a small group of companies, technological cooperation has been a strategic means of gaining and retaining competitiveness. We are speaking of top companies, big or small, which are known to spend considerable amounts of money on research. The case of Blokit, a small company which makes reagents, is well-known.

Other companies have begun to cooperate in part because of the impetus brought about by Spanish entry into the EC. This broadening of horizons allowed them to take part in European programmes. In the early stages, the first Framework Programme was near completion and it was necessary to wait.

Other companies have chosen not to cooperate. They have retained the classic relations involved in subcontracting. There is the feeling that "cooperating" sooner or later means losing independence; or else having to take part in European programmes. This last option which means having to deal with the Government, is seen as having to enter official, bureaucratic circles.

Action by Central and Regional Government

On the question of participation in European programmes we can see that the central government has acted as coordinator and catalyst. In general, however, the level of promotion of the industrial and development policies for which the regional governments are responsible is quite high. The decentralized state which emerged from the 1978 Constitution has been strengthened more and more in the last ten years. The autonomous regions have authority in certain area of industrial and technological promotion.

In practice, therefore, the Ministry of Industry has no network of centres around the state. It has promotion organizations such as the IMPI and the different regions have support organizations which have different names and activities. These organizations carry out their own activities and act as representatives for financing and assistance from central government. In the Mediterranean regions, for instance, we could point to the following organizations: CIDEM (Centre for development and business information) in Catalonia; and IMPIVA (Institution for the small and medium-sized enterprise of Valencia) in the Valencian region.

CIDEM had nearly 8,000 consultations in 1988, mainly with SMEs. The centre deals with many different topics including financing, regional incentives, foreign investment, R&D, industrial property, and such like. To be precise, its division for industrial links with Europe provides the means for Catalan companies to contact companies from the EC, promoting and encouraging technological and industrial cooperation. Thus, among the initiatives taken to provide information on current programmes we can find: the database EUROCONTACT; specific informative activities for the programmes ESPRIT, RACE and EUREKA; and in coordination with the CDTI (Organization of the Ministry of Industry) the distribution of information on CERN's public biddings for industrial components and high technology in order to ensure the participation of regional firms take part in them.

In addition, CIDEM offers also a "Euro Information Service" (the "Euroguichets" of the former Taskforce on SMEs now DGXXIII) on all kinds of EC activities of interest to the SME especially loans and financial support; how to adapt Spanish legislation to EEC legislation; how to find industrial partners; technical norms and certification international trade.

The OTT (Centres for theTransfer of Technology) and their Mission.

This global growth in cooperation has also affected university-enterprise collaboration. There was an increase in relations which it was found necessary to strengthen as much as possible; this led to the consolidation, in the universities, of the so-called "Centres for the Transfer of Technology". The OTT were created to encourage relations between universities and firms, but, meantime, the R&D National Plan Secretariat has begun to coordinate efforts so that the OTT can also act to support the participation of different university teams and public research organizations in European programmes, especially those which started in 1989.

Final Remarks

Spain's entry into the EC has almost coincided with the initiation of an "ambitious" National R&D Plan. One of its main objectives is to double GERD (Gross Expenditure on R & D) as a percentage of GDP, from 0.5% to 1% in a few years. In a parallel way, the action to encourage the presence of companies and organizations in European programmes has increased.

For instance, we find several Spanish companies from different sectors and of differing importance participating in EUREKA. It may be true that they are the top, dynamic companies, but it is also true that this significant presence and the effort that the Government has put into encouraging it must be seen as an attempt to:

- get a significant Spanish presence in as many strategic projectsfor future technology as possible and to avoid being left out of both formal and informal circles of exchanges of technology and knowledge which, in the near future, may become decisive.
- profit from these agreements so as to strengthen a group of dynamic and innovative firms which could act as an example to existing companies. It is no accident that the most successful companies taking part in EUREKA and ESPRIT, for example, are also those appearing regularly in the economic press.

Unfortunately, it is still too soon to estimate the real contribution Spanish companies have made to these projects. Meanwhile it must be remembered that the Government's concern has been justified and that, for the moment, this general approach should be maintained.

The coordination and diffusion efforts will in fact have to continue to increase so as to convince a larger number of competitive firms that cooperation is growing more and more essential in an economy in which markets are changing quickly, where strategic behaviour is also changing and, above all, internationalization is becoming irreversible.

Internationalization and Cooperation – The Irish Policy Response

D. O'Doherty

The currency tendency toward internationalization and cooperation has received a mixed policy response in Ireland. While "partnership" has become a widely recognized issue, it has still not been put at the centre of a new industrial strategy. May of the older approaches, especially towards the attraction of foreign direct investment (which in many ways were a pragmatic response to the last phase of internationalization and globalization from the mid '60s to the mid '70s), still continue, largely because of their success in the past, although a set of quite novel measures has been grafted on to the incentive and support system.

The relative openness of the country in economic, industrial and technological terms has meant however that the impacts of changes internationally have been transmitted fairly rapidly to virtually all institutions – industry, public research bodies, the state development agencies and even the universities to some extent even though the response in terms of policy has been somewhat slow. This openness is clearly a potential advantage for small firms and small countries if it can be used to introduce flexible integrated, strategically defined approaches to development. In the Irish case, elements of such an approach have been put in place but the challenge now is not so much to introduce additional measures but to pull together a "total innovation" and "partnership-based" incentive and support system.

In this context, the most significant new approaches which have been introduced include:

Company Development

In order to enhance the development of indigenous firms a "Company Development Programme" has been instituted with the emphasis on preparing "Company Development Plans" with a three-year perspective. This is intended to review the firm's existing competitive situation and its future prospects and to identify the best actions and state assistance to reinforce its strengths and reduce its weaknesses. The exercise is carried out by the firm's management, assisted by executives of the state technological and industrial development agencies and private sector consultants, as appropriate. The aim is to encourage management to examine all aspects of their business and to prepare a formal Action Plan.

The approach places major emphasis on the marketing and technological issues involved in product and process development, including the opportunities presented by technology transfer and other cooperation links, and on the financial, managerial and other key aspects of the business.

State agencies for all area of the firm's activities, including research and technology transfer, is now governed by the priorities emerging from this exercise.

Sub-supply by Small Firms

The state development agencies have come together on a task force basis to organize the development and implementation of a National Linkage Programme (NLP) with the following elements:

- a comprehensive data base on existing products and production capabilities as well as on the needs and requirements of MNEs
- selective promotion and development of "winners"
- co-ordination of state support and assistance for marketing, product and process development and quality control during the critical initial business development period

It is organized to operate simultaneously on three levels designated as Linkage 1, Linkage 2 and Linkage 3. The programme's main objective is to optimize the amount of raw materials, components and services sourced in Ireland by overseas companies. In particular it aims to develop trade link between the multinational companies who have established in Ireland and Irish based companies who could act as suppliers.

Linkage 3 in particular involves an explicit cooperation and technology acquisition element whereby the Irish firms involved will transfer in the technology they need to meet the requirements of their multinational customers. In addition, the subcontracting process itself leads to a transfer of existing technology through, for example, the design and adaption of products, manufacturing processes, and the organization and management of the firm. This can take several forms:

- licensing agreements for patents and know-how between the partners involved
- sale or transfer of second-hand machinery
- technical assistance
- quality control and monitoring of standards

- occupational training
- access to technical and other information

The aim is for the subcontractors to achieve a level of quality, specialization and overall excellence which will ensure that the principal will itself become somewhat dependent on the sub-supplier. The National Linkage Programme has achieved substantial success not only in generating new business but in raising the level of business planning and also the understanding of the innovation and partnership process in the participating companies.

Technology Acquisition

In response to the changes in the pattern of international movements in technology and commerce, a scheme of *Technology Acquisition Grants* (TAGs) has been introduced.

Grant assistance is offered towards directly incurred costs on:

- Licensor charges – patent rights, technical assistance, prototype components and licence options in the form of down payments, royalties etc.
- Licensee costs – wages and salaries, travel and subsistence
- Consultancy fees – specialist advice on the technology, patents or copyright

Assistance is available up to a level of 50%, subject to a maximum of £250,000 per project.

The project must involve a reasonable degree of innovation relative the company's existing level of technology

- The project must have a reasonable prospect of commercial success
- The product/process must be produced or applied within the State
- The parties to the licence agreement must be independent entities
- The company must provide a brief history and an indication of their Product Development plans as well as an assessment of the commercial and technical viability of the project

In addition to the TAG's scheme the state has a number of agencies involved in promoting and servicing the licensing and joint venture aspects of technology transfer at both a national and a regional level. There are also a number of private sector intermediaries operating in this field. The role of both public and private intermediaries is also promoted by the European Community's "SPRINT Programme". A 1987 SPRINT survey indicated (on the part of 80 per cent + of the respondents) a more widely felt need in Irish firms for "information and advice on potential partners" than in any of the other 8 member-states surveyed[1].

Research Collaboration

Side by side with the evolution of the EC Framework Programme for "precompetitive research" into a major instrument of Community policy – both in financial and in science and technology terms – the issue of local but particularly transnational collaboration on research problems and projects has surfaced in Ireland. Industrial R&D, however, still tends to take place almost exclusively within individual small firms, with state assistance being provided through an R&D Grants Scheme which funds up to 50 per cent of costs directly incurred. The overall level of Business Enterprise R&D also tends to be relatively low (Chart 1), although rising rapidly.

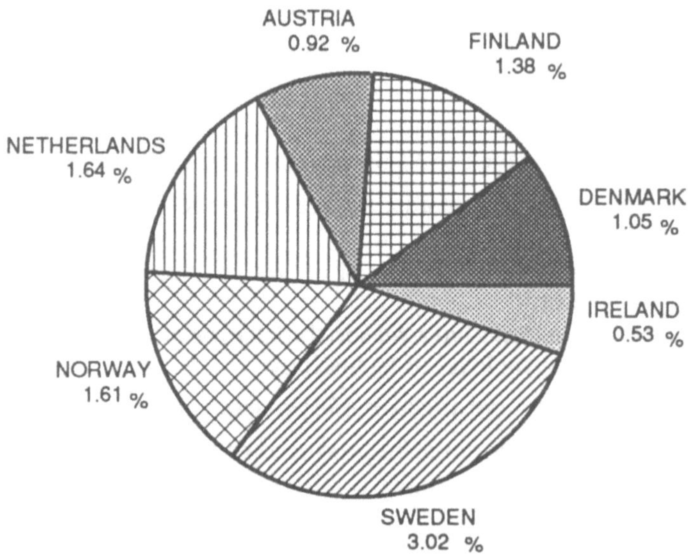

Chart 1: Inter.context: business sector R&D 1986
Domestic Product (%)

1 "Enquiry on Technology Transfer and Transnational Collaborations" – EC SPRINT Prog. DGXIII Luxembourg 1987

The Framework Programme, by treating collaboration and cooperation as an integral aspect of all projects, has however reinforced an emerging tendency within Irish industrial and technology policy towards more emphasis on firms jointly undertaking or contracting-out projects which address some common problem or theme. Such projects are still comparatively rare, with a severe lack of the associations and "clubs" which have come to characterize the research system in many other countries. The Framework Programme does, however, appear to significantly extend the participation firm's formal and informal networks of contacts and cooperation.

European Markets in the 1990s – the Irish Export Board's Strategy for Catalonia

L. Kelleher

In the past 18 months business in Ireland has become much more aware of the creation of the Single European Market. The programme is already far advanced. The process of removing physical, technical and fiscal barriers is scheduled for completion by December 31st 1992.

Important though these changes are, its clear that much more important for companies is the attitudinal and behavioural change which is sweeping business across Europe, in the expectation that the Single European Market represents something qualitatively different. This expectation is causing companies to re-appraise their market strategies.

The 1992 changes are affecting different sectors and different markets in many ways. The removal of barriers, advances in information technology, and improvements in transportation and distribution, will "shrink" the geographical limits of the market, and override the present dominance of national geographical boundaries.

The underlying belief is that the creation of the Single European Market is having an impact far beyond the removal of the identified barriers: it is shaking up markets, causing companies to re-appraise their market strategies, and to develop new strategies. The mental picture for a company must be: how can you build long term business relationships in existing and new markets? How do you create a situation so that you are an insider, marketing within European markets, rather than someone on the fringe selling from a distance into a European market? (and thus mentally both to yourself and to your customer, a fringe element).

Anticipating these changes, Córás Tráchtála is this year launching a number of new strategies. A key one of these is a Regional Development Programme.

This year the programme focuses on two growth regions, Lombardy in Italy and Catalonia in Spain. The objective is business development through identifying and cementing suitable marketing partners for Irish firms. We have chosen these regions because they are fast growing, outward looking, European in their orientation, and with an existing exposure to Irish goods and services. They are two regions which we feel will gain disproportionately from the completion of the Single Market.

Catalonia is Spain's foremost economic region accounting for:

- 16% of the total Spanish population

- 19% of the total Spanish GDP
- 25% of the total Spanish industrial GDP

Catalan income per capita is more than one third above the Spanish average.

The region has a large degree of autonomy, with its own government or Generalitat. The Generalitat has a wide range of responsibility in commerce, consumer and environmental matters.

Catalonia has an extensive road, rail, sea and air transport network with road and rail links to the equivalent French systems, as well as with the rest of Spain.

Catalonia has a highly diversified industrial structure, and is particularly strong in the sub-sectors of metal goods, chemicals, paper textiles and food processing.

The specialization in processing as opposed to the heavier basic industry has the advantages of greater flexibility and adaptability, and a high growth potential.

The region has a population of 6 million people. The dominant city is Barcelona with a population of almost 2 million and the major industrial and commercial centre for the region. The zone around Barcelona contains 80% of Catalan industrial jobs, 80% of the industrial value added, and 78% of total Catalan production.

Key elements in our programme for Catalonia are:

1. A marketing programme aimed at companies in Catalonia, with the purpose of enhancing business development between Ireland and the region. The programme is initially broad brush, and then quickly focuses on targeted sectors and categories of company. The first screening was undertaken in May/June 1989.
2. A top quality promotional campaign, integrated with the marketing programme commenced in May.
3. Participation at the Barcelona Trade Fair, 6 - 11 June 1989.
4. Combining a business group visit from Catalonia with a "Spanish Buyer Airlift" on 21 and 22 June 1989.
5. A second phase marketing programme, aimed at specific sectors of companies expanding within the European Community.
6. Business development missions to and from Catalonia, in Autumn 1989.

The focus of these programmes is market pull, rather than push out of Ireland: we are seeking to create a demand in Catalonia, from Catalan firms, for marketing alliances with Irish companies. These alliances, for an Irish firm, may take the form of:

- links with an importer/distributor, who provides marketing and product development support.
- links with a design/product development company, strong in a particular market segment.
- links with a large purchaser, who sets quality standards, distribution standards, market service requirements and product development ideas.
- links with a competitive or complementary manufacturer having established market knowledge and established market presence, where product lines are at least partly complementary.
- looking for a marketing alliance which strengthens the Irish company in another market area of the European Community as well as Catalonia.

In the programme, an important element is the development of links with "business services" organizations. These include representatives of Chambers of Commerce, industry associations, regional development organizations and key members of the press – all important opinion formers on behalf of the programme in the Catalan region.

International Company Development Programme

T. McGuinness

Introduction

The idea for an integrated company development programme through a transnational approach was originally conceived by the Irish Productivity Centre three years ago. Based on its experience over many years working with small and medium sized businesses in Ireland, the Centre was aware of the difficulties growing enterprises had in assessing their strengths and weaknesses, developing management/leader group skills, introducing technology changes and realizing their export/market potential.

The Centre approached the Jutland Institute of Technology, Denmark and C.I.M.K. in the Netherlands, both Institutes working with small and medium enterprises in their respective countries, with the idea. While the initial concept was well received, some refinement was necessary to take account of the different cultures and needs at company and national levels. After a number of meetings between the institutes, the "International Company Development Programme" was designed and submitted to the EC under the "Innovatory Pilot Project" scheme and approved for funding.

Programme Design and Implementation

The Programme was aimed at companies preparing for an international future, who had a track record in the home market, employing between 50 and 150 people and had sufficient differentiation and expertise at management level to make inclusion in the programme feasible and worthwhile. Six companies from each country were selected for participation.

The Programme set out to develop the management teams of the participating companies, as well as increasing their selling opportunities in Continental Europe, particularly in the context of 1992. The companies selected were seen as "open" to increased international marketing and consequently, the Programme aimed to help them achieve "critical mass" levels of experience, knowledge, expertise, resources and relationships necessary for successful international marketing. Achieving this "critical mass" is very difficult in a small domestic market or by purely exporting to "close" countries such as the UK in the case of Ireland.

The Programme also had as a by-product the development of a three-country design for assisting SMEs which brought together the experiences and approaches undergone separately in each country and applied them in a relatively uniform way. The central thrust of the Programme can be described as being focused on the development of the management team in an international context. The design of the Programme was based on a combination of

- Formal classroom training sessions
- Group work of two types
  i. related to particular disciplines
  ii. related to company leader groups
- Presentations by sector specialists
- Company counselling
- Organized visits to other companies

During the programme, which extended over 15 months (February 1988 to June 1989), six national and three international seminars were held, together with nine one-day meetings between individual companies and their counsellor.

National Seminars

Six two-day seminars were held within each country throughout the period. Each participating organization nominated its executive/core group, which included not only the Managing Director but individuals from the various functional divisions within the company - sales/marketing, technology/production and finance. Individuals from each discipline of the business were provided with an opportunity not only to analyze the strengths and weaknesses of their own areas of expertise but also to gain an insight into the operation of the company as a whole. By having the full management team involved, every facet within the company critical to its international success was addressed to give the company a strong coordinated entry into their increasingly international future. The themes of the national seminars centred around the following.

- General Management
- Export Marketing Management
- Financial Control Management
- Technology and Production Management

- Leadership and Organization
- Strategic Management and Planning.

Within this "workshop" environment, the management team in each company had the opportunity to work together on specific projects which were of immediate concern to them, whether financial, marketing or production based, to come to an agreement on fundamental matters and design a strategy for implementing such changes/projects.

Various training methodologies were used at each seminar, including lectures, presentations from specialists, group work, videos, and role play exercises.

International Seminars

Three international 3-day seminars were also held in the period, one in each of the participating countries. The three themes for these international seminars were:

- Export Marketing Management (held in Amsterdam)
- Technology and Production Management (held in Arhus, Denmark)
- Leadership and Organization Culture (held in Dublin).

These were attended by the Managing Director and the manager from the production, sales or finance, depending on the content of the particular seminar, from all 18 participating companies.

The purpose of the international seminars was to provide a better understanding of general and international business through intermingling and cross-fertilization between the companies. Participants gained first-hand experience in international negotiation and cooperation and had the opportunity to explore cultural and communications difficulties in a non-threatening environment, thus breaking down many of the barriers which otherwise would have prevailed.

A series of company visits in the host country were organized during the international exchange and as a result of this some companies have made valuable trading links which will be developed on an on-going basis.

Counselling

Throughout the programme each company was visited by a specialist for counselling sessions. This was an integral part of the design of the programme. Each module was supplemented by individual counselling for the core/leader group in their own environment at company level. This "in-house" training was designed to identify and help them with any problems experienced in the preparation and implementation of their strategic plans. The counselling also enabled the participants to translate into practice the contents of the Programme under the supervision of an expert.

An Overview

The International Company Development Programme was generally regarded by the participants as being successful overall. Some aspects emerged as being highly significant and as contributing to the successful outcome of the project:

- the involvement of three countries in the design and implementation of the programme ensured that trainers and facilitators brought their own distinctive cultural and national competencies to the project work - thus exploiting and using as strengths those differences which are frequently seen as a constraint.
- the transnational structure underpinned the concept of shared learning and shared experience at all stages of the project for both trainers and participants;
- the internationalizing of Irish SMEs, particularly in the European context, was seen by the companies as the only way for survival and growth in the long term;
- the demand on participating companies for total commitment to the project and its objectives;
- with all members of the management team involved within each SME a considerable cross-disciplinary element emerged in the training which appears to have a number of advantages over the single discipline training and up-dating that is often a feature of Irish companies in this size category.
- a great deal of emphasis was placed on building a sound training programme throughout the project. The basis of this process was the active involvement of each participant in working on the development of his or her own company in a constructive environment.
- the companies had a clear view of their own strategic objectives in relation to their development in an international context. In addition to this, they also gained valuable experience in implementing change in their own companies and applying external knowledge to their particular situation.

Business Development Through Cooperative Agreements – an Innovative Approach

A. Duke

Background

The need for Irish industry to develop new business and markets has been well articulated, as has the shortage of expertise and skilled personnel to identify and develop opportunities in the home and export markets. Small and medium-sized businesses in particular need to adopt and adapt new technologies and techniques if they intend to remain competitive. Low profit margins and reserves play a significant role in postponing development and investment programmes as companies cannot afford the risk of project failure.

The business cooperation process, whether it be licences, joint ventures, agency, distribution or other agreements, offers the most comprehensive solution to the dilemma of the small and medium sized company. However, the shortage of personnel skilled in these areas, and lack of awareness of the rewards of interbusiness cooperative agreements has resulted in the full benefits of this process not being achieved.

A programme has been designed to provide training in the cooperation approach to selected participants who have then been seconded to suitable companies. These participants completed a business development programme for the company in which they were placed, extending over a 30 week period.

The promoters of the Programme were:

- FÁS – the Training & Employment Authority (Republic of Ireland)
- DED – the Department of Economic Development (Northern Ireland)

The objectives of the programme were:

- to train participants in the methods and processes of business development utilizing the approach of International Business Cooperation.
- to provide a "vehicle" for SMEs both in the Republic of Ireland and Northern Ireland to research, identify, evaluate and negotiate business development opportunities utilizing the "cooperation route".

Eolas, The Irish Science and Technology Agency, managed and operated the Programme, while outside inputs were provided by consultants such as Technology Link International, a firm specializing in US/Europe Cooperation and other US and European-based specialists.

North America was chosen as the "target" area to search and identify the product/technology opportunities which would be compatible with the development requirements of the Irish companies.

Eleven participants were selected for training, each of whom had senior to middle management experience. These in turn undertook a cooperation assignment in each of the eleven companies selected for programme inclusion. A senior company executive (usually the Managing Director) was closely involved at all times. The companies were drawn from a number of industrial sectors:

- Specialized Software
- Precision Engineering
- Heavy Engineering
- Water Treatment
- Food Processing
- Horticulture-related
- Construction-related Products

The Key stages of the Programme were as follows:

- Company Appraisal
- Business Development Search Brief
- Research
- Contact
- Contact Response Evaluation/Selection
- Overseas Itinerary Planning
- Overseas Visits
- Evaluation of Opportunities Sourced
- Action Plan for Business Development

At all times the training in the techniques and methodology of International Cooperation and the utilization of that training to implement the company assignment operated in tandem.

The choice of Cooperation route was wide, including:

- Licensing
- Technology Transfer

- Joint Venture
- Distribution
- Joint Marketing

or combinations of the above.

The strategy was to provide each company with:

- a large number of potential opportunities identified from U.S. corporations seeking some type of cooperation arrangement;
- structured evaluation to screen these opportunities into a short-list of what was of most interest to the company;
- arrangement for the participant (on the company's behalf) to visit these short-listed corporations to access further information and assess the potential of the opportunities presented;
- assistance to the company in further evaluating the opportunities presented, in order to select the corporate partner(s) which would best satisfy development requirements.

Each company (assisted by the participant and programme organizers) undertook an appraisal of the existing business. From this each company decided on the area that they considered offered the most potential for business development. In most cases this involved some form of diversification.

The participant then undertook detailed research to identify U.S. corporations operating within this selected area.

Research included:

- Hard Copy Research Material
- On Line DataBanks
- Trade Organisations
- Research Institutes
- Specialised Publications, etc.

There was an overall average of 300 corporations identified for further contact per programme company.

A contact letter was sent by the company to these corporations outlining their requirements and the benefit of cooperation to the U.S. corporation. This letter followed a set format aimed at achieving maximum result.

An overall programme average of 25% positive responses resulted from this contact process.

From these responses, each company evaluated the potential vis a vis their requirements, and selected a short-list of corporations which would be visited by the participant on their behalf. An overall programme average of 15-20 corporations per company was selected and visited.

Participants travelled to the US for a period of six weeks to meet with and evaluate further the potential for cooperation in all cases these meetings were with senior executives, and in most cases with the corporation President. These corporations were spread throughout the US Full support services were provided in the US at all times during this phase by the programme organizers and coordinating consultants.

Following return from the US, detailed evaluation of the opportunities sourced through the programme took place, and the necessary follow-up actions were completed.

By the end of the programme the following results has been achieved:

- a number of International Business Agreements had been signed and were being implemented;
- market research was underway by a number of companies to ascertain market reaction to and the size for sourced opportunities;
- a number of US executives had already or were in the process of visiting Irish companies for further discussion;
- a number of Irish Managing Directors were scheduled to visit US plants for further discussion.

The final number of International Cooperation Agreements and their impact resulting from the Programme should be measurable within a 12 month period from the Programme close. In conclusion, the Programme achieved its objectives and exceeded expectations. A number of factors can be identified as contributed to it's success, including:

- design and methodology: by the training an individual in the process of International Business Cooperation and the integration of the individual into a company to undertake a structured cooperation "search" assignment, with the back-up of the assembled expertise this programme provided, ensured a success many companies could not hope to achieve from their own resources.
- the commitment of both company and individual. By its very nature a search assignment in this area demands a high degree of commitment, flexibility, professionalism, and hard work by all concerned.
- the "background" nature of the support. The individual participant became part of the company, and the company was the only "public

face" seen. He or she instigated the contact, met with the US corporation, and conducted the evaluation processes. The support and expertise provided by the Programme was at all times present but was never seen other than by the programme participants and companies. Intermediaries were not used.

Section 5: Conclusions for Research and Policy

The Importance of the Cooperation Phenomenon

Cooperation and networking between firms is becoming increasingly important whether it involves small or large firms or whether they are supplying, manufacturing or purchasing goods and services. There are many reasons for believing that this is also a permanent phenomenon, with very important long term economic and technological consequences.

Cooperation and networking arrangements are motivated principally by the need or desire for technology, management and skills exchanges; by the need to share risks and the need to enter new markets. In the long term both the firms directly involved and the domestic economy must benefit from successful well-managed cooperation arrangements because their portfolios of technology and expertise is extended and the overall technological and other risks are shared and reduced.

But many factors influence the form and success of cooperation, and there is a complex interaction between the ways in which firms cooperate and collaborate and the economic and technological effects of this cooperation and collaboration. It is evident that there are a set of inherent structural weaknesses facing the smaller firm and smaller economy which can hinder effective and efficient cooperation. These require special study and specific policy action.

Given the structural weaknesses at national level outlined above, much of this must be at an international level. It is also clear that such actions must take place within broader technological and industrial policies.

The OECD and the European Community – Small firms and small country issues

While much still needs to be done to ensure that small countries, often with "late entrant" disadvantages, are in a position not to lose out in (whatever about gaining from) the current globalization process, some of the necessary means are being put in place. We have seen that there have been some moves at the Community and at the wider European level, for example through the EUREKA programme, to recognise that there are specific small firm/small country issues. The EUREKA Secretariat has carried out a major unpublished study which has identified both "structural" and "programme" barriers to SME participation in projects and, as we have seen from the Secretary General, in principle at least is trying to reduce them. Irish participation, however, as shown in Mr Meyer's paper, continues to be low.

The Technology and Economy Programme (TEP)

There are obviously still many unknowns in this area. OECD is one organization working to fill the knowledge and policy gaps, for example in the framework of the Technology/Economy Programme (TEP) launched in 1988. This is aimed at examining the complex relationships between technological, industrial and economic development.

The cooperation and networking phenomenon is one focus of the TEP activities and there will be a chapter on networking in the final TEP report which is due in mid-1991.

The policy conclusions which may be expected include the need for overall technology/industry policies which are based on a realistic assessment of the needs and market pressures on small firms, and specific measures designed to reinforce cooperation within those overall policies. These mirror the findings of this book.

Community Programmes and Policies

Similarly with most of the individual EC research programmes and with the overall Framework Programme itself: increasing acknowledgement of these specific factors and problems has been forthcoming but "competitiveness" objectives and "excellence" criteria are dominant and are seen (many believe wrongly) as limiting the scope for responding to other concerns. Small projects are, it is stated, being given an unspecified level of preference, and a "feasibility study" phase has been introduced in a small number of programmes notably BRITE/EURAM. Many commentators agree with Roy Rothwell (see later in this section) that only a separate programme or a prescriptive share of the existing programmes (like the Small Business Innovation Research (SBIR) programme in the US) will really make a difference for SMEs. (See also,for example, "The Irish Policy Response" in Section 4 above.) The Structural Funds, including proposed programmes such as STRIDE (Science and Technology for Research, Innovation and Development in Europe), administered by the Regional Policy Directorate General (DGXVI), are seen as being mainly infrastructural rather then project specific. However, it is important to highlight a number of novel approaches for SMEs which are currently being proposed and tested. These directly address the needs of the mass of smaller firms as yet untouched by Community science and technology programmes. They also retain the element of partnership and collaboration which has been characteristic of EC initiatives. They include:

– "CRAFT" – a programme to bring together groups of SMEs which themselves do not have the capability to carry out R&D but which with EC financial assistance will fund the carrying out of the work in a large firm or a research association in a Community country.

– a "subcontractors" programme which will aim directly at upgrading the technological capacity and role of subcontractors in the Community. Already subcontracting has lost much of its dependency image and a number of small technology-based Irish firms, for example, have preferred to adopt a subcontract-

ing strategy rather than act as full partners in ESPRIT and other research programmes.

The Community's MONITOR Programme which is concerned with the examination of European science and technology policy, is currently looking at both the conceptual and policy issues concerning small firms and small countries in a global technology and economic context.

Without ignoring the significance of national initiatives in identifying and tackling many of the small firm, small country, cooperation-related strategic issues raised in this book it is clear that the European Community (and so the Commission) have a major role to play in ensuring that these issues are tackled in the most coherent and integrated manner, within an overall European science, technology and industrial policy.

A number of the contributors to this book have stressed the lack of studies on small firms, small economies and the cooperation phenomenon, which would inform us on the kind of firm and sector specific issues we need to address in policy terms. A number have also stressed the need for cooperation to be seen in the context of overall innovation and technological change in small firms and small economies, and the policies addressing these processes.

That is why the book finishes with some specific conclusions and two papers setting out a broad agenda for European research and policy actions for small firms and small economies. It is only within such a context that the full potential of the cooperation phenomenon can be realised.

In this section three contributions are presented. The first summarizes the issues for research and policy arising from this book and the conclusions of the Dublin Conference for which most of the papers in this volume were first written. Here a major focus is on the need for further studies and the importance of good intermediation. The second draws on a submission made by Roy Rothwell to the Industrial Research and Development Advisory Committee (IRDAC) which advises the European Commission on science and technology matters. This deals with the requirements for and the implications of an integrated European Community "small firms" policy and stresses the need for a European SBA (Small Business Administration) as well as more specific policies, particularly for encouraging small firm - large firm interactions. The final paper is a proposed list of issues/studies which address the emerging small country/economy issues and within which cooperation strategies and policies at the European level in particular have an increasing significance.

Small Firms, Small Economies and the Cooperation Phenomenon – Issues for Research and Policy

D O'Doherty

In this note the issues arising from the papers presented in this book (as well as from the Cooperation Phenomenon Conference in Dublin) for research and policy are summarized. All of them are put forward within universal agreement that a great deal more attention must be paid to the characteristics and roles of small firms and small countries in the cooperation field.

Variety

There is a broad consensus that cooperation is characterised by a variety of participants, cooperation and geographical arrangements and firm strategies. Cooperation is a complex phenomenon which is best understood and best undertaken within and "overall systems" or "total innovation" approach by the firm.

Studies

There is need for explicit funding of studies, at the national, European Community and OECD levels on:

- Cases/Success Factors – there is need for broadly-based studies of "partnership" arrangements, but also for more in-depth research at the individual firm level to define needs and objectives in terms of what constitutes success in cooperation.
- Size issues – the specific needs and advantages/disadvantages of SMEs in R&D, licensing, market cooperation etc. and the significance of informal and personal networks which appear to have a major role.
- Cross-country surveys – there is a need to analyse different national experiences in cooperation, particularly of small countries, against the wider economic, social, and particularly cultural backdrop.
- Policy impacts – the design and impact of regional, national and international policies should be analyzed and evaluated. The larger countries would appear to have more experience and commitment in this regard.
- Socio-cultural factors – these would appear to play a decisive role in the nature and success of cooperation and networking activities and different "national systems" of innovation seem to be both charac-

teristic of and highly significant for small countries and their small firms.

Bilateral Cooperation

An explicit "interfirm" element should be introduced into the more classical science and technology cooperation between individual countries. There is evidence that this is currently taking place. There is also a need to compare and contrast policies and programmes for encouraging interfirm cooperation on a continuing basis.

Actions

National and international action, particularly at the E.C. level, should be taken to encourage a wider diversity of both public and private intermediaries. This emerges as one of the most important conclusions from examining the "cooperation phenomenon". There is also a need to deepen mediation skills while encouraging such diversity. Direct knowledge of local conditions and of the parties involved is essential for good mediation between firms, particularly at the international level. Intermediaries should offer help in negotiations and in the preparation of model agreements. At the same time it is important that management remains directly involved in the cooperation process at all times.

Generation/diffusion

For small firms and small countries cooperation arrangements aimed at diffusing technology are at least as important as the R&D agreements which have to date been the major focus of research and policy at both international and national levels.

SME Forum

The issues raised in this book need to be discussed at the highest level in the OECD, European Community and national contexts. At the Dublin Conference, a committee on the lines of IRDAC, which advises the European Commission on science and technology, was suggested for small firms. This would take up the specific small firms issues in conjunction with IRDAC.

Specific Policies

There is a consensus on the need for specific policies at all levels aimed at encouraging SME cooperation. However it is felt that firms should not be over subsidized or pushed into agreements which do not fit with their overall business strategies.

Summary and Policy Recommendations to EC-IRDAC Working Party on "SMEs and their Role in New Technologies"

R. Rothwell

Introduction

It is my belief that SMEs in Europe are now playing a much more significant role in innovation than they were a decade ago. The available evidence on this point is, to be sure, fragmentary, but I think convincing, and it provides me, at least, with considerable cause for optimism. I will begin by summarizing part of this evidence.

Appendix 1 to this paper presents firm size/innovation share data for the UK. It plots, for eight different time periods between 1945 and 1983, the share of *important* innovations introduced by British firms in five firm-size categories. The data are taken from the SPRU Innovation Databank which contains details on 4,500 significant British innovations. Perhaps the most notable feature of the data is the increasing share of national innovations taken by firms in the size category 1-199 employees. For SMEs (employment between 1-499) the innovation share increased from 22.6% during 1965-69 to 29.2% during 1975-79, to 38.3% during 1980-83. So, clearly, something significant is occurring in the UK SME sector, at least as far as national innovation share is concerned.

The second set of data relate specifically to new technology-based firms (NTBFs). Appendix 2 to this paper summarizes the results of a study published by Arthur D. Little in 1977 which attempted to determine the number of NTBFs in existence in the UK and the FRG in 1975 that were formed between 1950 and 1975. The figures (albeit probably underestimates) indicate a low level of NTBF activity in the two countries both in terms of number of companies and their economic impact.

Appendix 3 to this paper summarizes the results of a more recent, follow-up study conducted by Segal Quince and Wickstead of Cambridge (UK) and the ISI Institute of Karlsruhe (FRG). It indicates a massive increase in the number of NTBFs (including software companies) in both countries. Annual NTBF start-up rates have been increasing rapidly since 1979 and the number of NTBFs founded in the FRG in 1986 is estimated at nearly 500. Despite this recent increase in NTBF formation in the UK and the FRG, it is likely that relatively few of these firms will grow to achieve large firm status.

One reason for the dramatic increase in NTBF formation in Europe since the early 1970s is the nature of several of the emerging "technology clusters". (These are shown in Appendix 4 to this paper.) In the information technology and biotechnology areas, entry costs for new firms appear to be relatively low. In these areas innovation is *knowledge intensive,* rather than capital and scale intensive; and information technology in particular has opened up many new market niches suitable for entry by small firms.

This point can be illustrated by considering the case of electronic computers in the UK. During the 1950s and 1960s, UK production was composed almost entirely of mainframe computers, the introduction of which involved very high R&D, capital and servicing costs. During this period, the SPRU innovation data show that UK SMEs produced very few innovations in the computer industry. Following the advent of the integrated circuit but more especially of the microprocessor, entry costs declined dramatically and many new market niches opened up for computers and peripherals: SMEs' innovation share in the UK grow rapidly; it was 36% by 1970-74 and reached 64% during 1980/83.

A second factor contributing to the growth of technologically innovative SMEs in Europe during the past decade has been the rapid increase, since the early 1970s, in the number of public policy instruments introduced that are directed towards SMEs in general. Since 1980, throughout Europe, many schemes have been introduced to stimulate specifically NTBF formation and growth.

Policy Implication of the IRDAC Study

NTBF formation (and growth) on an appreciable scale has long been regarded as a uniquely US phenomenon. In this section, I will look at those factors which are generally understood to have been favourable to NTBF formation in the USA and attempt to determine to what extent Europe is, or is not, moving in the same direction and I will attempt this specifically in the light of the results of the IRDAC study. This might provide some indication regarding possible useful policy initiatives by the European Commission. Factors favourable to NTBF formation and growth in the USA are listed in Appendix 5 to this paper. These will be considered separately below.

The Home Market

I do not believe that there is any lack of "innovation-demand" in the national markets of Europe. This is clearly illustrated by the large volume of high technology goods imported into the EC member states. There exists, however, the major problem of a lack of market integration in Europe. We still do not have a "common" European market, with common standards and regulations, and non-tariff barriers to trade persist. Clearly, this is a prime area for action by the Commission. While comprehensive action is proposed, I hope that sufficient number of European NTBFs survive to take advantage of the "common market" intended for 1993!

Culture and Attitudes of Society

Societal attitudes are, I believe, changing better towards favouring wealth creation by manufacturing and related firms in Europe. During the 1960s and 1970s, industrial production was often equated with:

- the exploitation of workers
- pollution generation
- the general despoliation of the environment and so on.

During the 1980s, industry increasingly is seen as being associated with:

- Employment creation
- Wealth creation to fund social services

The "new" attitudes are especially associated with small firms and the notion that "small is beautiful" is now widespread in Europe. So, attitudes in society do appear to be changing positively, and entrepreneurs increasingly are being seen as valuable members of an economically dynamic society. In my own country, the term "enterprise culture" is heard more and more often.

Mobility of Individuals

In the USA, entrepreneurship is strongly associated with the *displacement* effect. Entrepreneurs were often those who were dissatisfied with large firm employment or who were made redundant, or they were immigrants or refugees. During the past decade we have seen strong displacement effects in Europe, with company closures, reductions in real expenditure on academic research and redundancies caused by firms' rationalization and de-diversification activities.

There has also been some movement recently towards increasing the mobility of pension rights (at least in the UK – I am not sure how widespread this move is) which effectively reduces the financial risks for spin-off entrepreneurs. In the specific case of the UK software industry, job-hopping and spin-off appear to be the rule rather than the exception.

And, at the same time, a number of large European companies are taking positive steps to assist SMEs in general and spin-offs in particular.

So, overall, I believe that job mobility and spin-off may very well be on the increase in Europe. I am not at all clear that there is a strong role for the Commission to play in this area.

Belief in the Importance of SMEs

There seems little doubt that to-day throughout Europe there exists a strong belief in the economic, technological and social importance of SMEs. This is reflected in both national and EC public policy statements and in the promulgation of many measures to assist SMEs. We might, however, pose the question:

"Would the creation of a European Small Business Administration, along the lines of the US SBA, represent a useful initiative for the EC?"

It would, I believe, be important from a *symbolic* point of view, if nothing else. A number of the IRDAC case studies indicate that SME managers perceive EC high technology initiatives to be designed mainly for, and by, large firms – so a major symbolic gesture might not go amiss.

Perhaps a European association of SMEs would suffice – a small firm European Industrial Research Managers Association (EIRMA), acting as a "marriage broker". Whatever form it takes, I do believe that we need an EC-wide, high profile linkage/support mechanism for SMEs.

Venture Capital

Up to 1980, there existed a marked paucity of venture capital throughout the EC. Since 1980, however, as Appendix 6 to this paper illustrates, venture capital funds have proliferated in Europe. This has most notably been the case in the UK, where two public policy initiatives played a key stimulating role. These are:

(a) the introduction of the Business Start-up Scheme (now the Business Expansion Scheme) which offers tax and capital gains concessions to investors in unquoted businesses; and

(b) the establishment of an Unlisted Securities Market in 1980.

In 1985, the total venture capital pool in Europe was about £4.75 billion, of which 40 percent was UK based. Currently, the US venture capital pool is about three times greater than the European. There has been a very considerable accumulation of management and evaluation know-how in the US venture capital industry; US venture capital is "hands on"; and there exist efficient inter-fund informal communication networks which effectively enhance the efficiency of the industry. It seems reasonable to suggest that, because of these factors, US venture capital currently operates more effectively than its European counterpart.

Despite the recent growth in venture capital in Europe, and for a variety of reasons, I believe that we do need an EC venture capital fund. I believe that the fund should be *technology*-based rather than industry-based. Given the *pervasive* nature of emerging technologies – and especially information technology – whose use spans many sectors, I think that a technology-orientation makes the most sense. A technology-based fund would allow for technological and market know-how accumulation in specific areas on the part of fund managers.

If there is to be an EC venture capital fund, it is my belief that it should be *hands-on*; offering management, financial control and marketing advice, in addition to finance. Management, manufacturing and financial control problems were crucially important during transitional growth phases in most of the companies studied on the IRDAC project.

While venture capital availability generally is increasing in Europe, there does still appear to be a shortage of *seed capital*. In 1984 and 1985, for example, the percentage of total national venture capital devoted to seeding start-ups was:

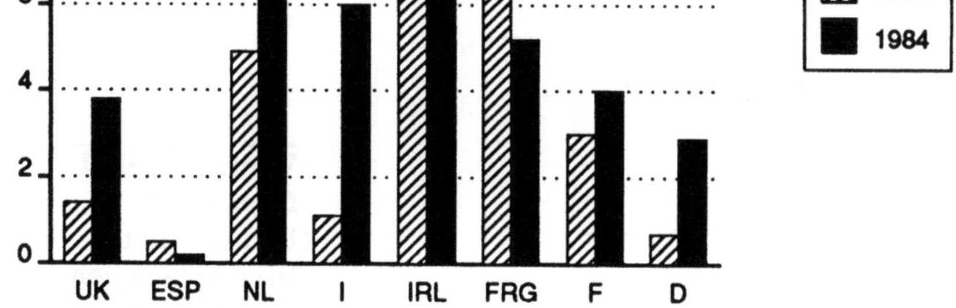

Chart 1: National Venture Capital (%) for seeding start-ups
(Source: Financial Times, December 8th, 1986)

Clearly, these percentages are small. I will return to this point in the next section.

DoD support and the SBIR

The role of US Department of Defence (DoD) R&D support in the emergence of, for example, the US semiconductor industry is well known. Both R&D funds and defence and space procurement played key roles in stimulating the emergence and rapid growth of NTBFs in this sector and others. More recently in the USA, we have seen the introduction (in 1982) of the Small Business Innovation Research scheme. Under the SBIR scheme, the twelve federal agencies which between them fund 99% of government R&D, solicit proposals from small technology firms. The scheme involves an increasing percentage of each agency's external R&D budget up to 1.25%. Commitment to SBIR between 1983-88 has totalled $1.5 billion.

SBIR operates in three phases:

Phase I: grants up to $50,000 in the first year for project evaluation.

Phase II: grants up to $500,000 for project development.

Phase III: final development and marketing which are funded from private sector sources.

Because firms entering Phase I & II are subjected to careful evaluation, demand from private venture capitalists to fund Phase III is high: they compete vigorously for SBIR firms. The ratio of private to public funds has averaged 8:1. Evaluations suggest that the SBIR scheme has been a considerable success.

It is my belief that an EC SBIR-type scheme, linked to an EC venture capital fund, would represent a particularly powerful combination, especially if the SBIR-type funds were directed towards *new* and *young* firms (to fill the seed capital gap).

Essentially, we would have:

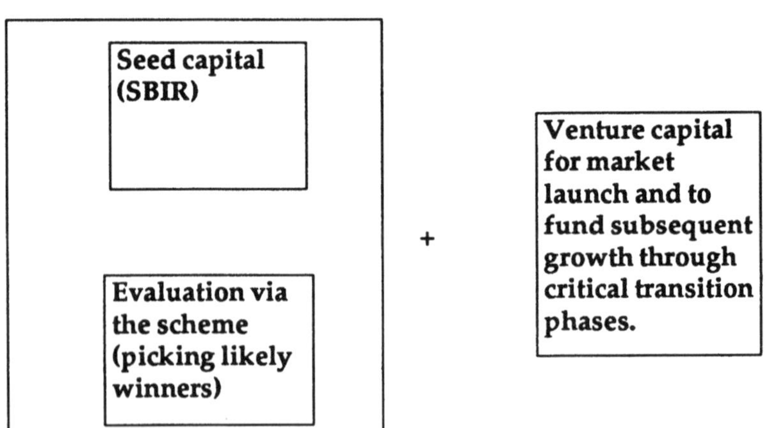

In fact, since firms both young *and* old suffered severe capital shortage problems during expansion phases, perhaps the term *development capital fund* would be more appropriate than venture capital fund since the latter tends to be associated specifically with NTBFs. The SBIR and development capital funds could be operated by the proposed European SBA – or by the European Agency for the Support of Business expansion proposed by some members of IRDAC.

Public Procurement

It is clear from a number of the IRDAC case studies that national and EC institutions could play a more active role in adopting purchasing policies – preferably innovation or quality-oriented purchasing policies – that provide easy access to SMEs. Some positive discrimination in favour of SMEs would certainly be helpful. It is worth noting that the SBA in the USA ensures that SMEs gain a "firm" share of public procurement contracts.

To digress slightly, in general most EC SME assistance measures operate on the supply-side. But we should not forget the importance of demand and in particular, from the point of view of innovation, the importance of producer-user linkages, especially between small and large companies. Such links have been very important to US SMEs. Any inter-company linkage schemes introduced by the Commission should include producer (small firm)-user (large firm) linkages in addition to R&D linkages.

Large/Small Firm Interactions

While it is clear that small and large firms interact symbiotically in a number of ways, it is equally clear from the IRDAC case studies that many European SMEs are reluctant, for a variety of reasons, to engage in *joint R&D ventures* with large firms. They do, however, wish to engage in joint R&D activities with other SMEs. Clearly the EC has an important information and linkage role to play in this respect. But I believe that the Commission should go further than this and initiate special technology programmes exclusively for SMEs. In other words, the Commission should establish SME joint R&D venture programmes in strategic technological fields in which SMEs can and do compete effectively. This would include the establishment of SME research clubs. They should be significant programmes in their own right and not be seen as marginal additions to programmes involving mainly large firms.

In addition I believe we should seriously consider initiating demonstration projects involving small/large firm joint R&D ventures. These would comprise a valuable cumulative learning process – establishing the "rules of the game" – as well as demonstrating mutual large/small firm benefits from collaboration.

University/Industry Linkages

It seems clear that, in contrast to even only five years ago, universities and industry throughout Europe are now forging much closer linkages. Since 1980 Europe has experienced a proliferation in the construction of campus-based science parks, innovation centres and incubator units.

I would be very surprised indeed if the entrepreneurial tendencies of academics in Europe were not already on the increase. I am not altogether sure what the Commission might do to enhance this trend – except, perhaps, to increase the supply of seed capital in the way I suggested earlier.

In terms of stimulating linkages between existing SMEs and universities, I think that schemes to subsidize SMEs to contract-out R&D to universities and other research institutions would be helpful. Such schemes have, I believe, already been established in France and the FRG.

In addition to the factors listed in Appendix 5, I would like briefly to address the issue of *skills and training*. It is clear from the case studies that many of the companies suffer crucial skill shortages and find it difficult to compete with large firms in the labour market for technical specialists.

At the same time it is equally clear that the firms need to access *external* technological expertise as a necessary *complement* to their in-house R&D activities. The point is, there exists strong evidence to show that an SME's ability to *access, transfer and assimilate* external technology depends to a large extent on its level of in-house technical expertise. Thus, technical skill shortages can disadvantage SMEs in two ways:

- first, inhibit in-house technological accumulation; and
- second, inhibit the transfer and assimilation of external technology.

I would like to suggest that the Commission consider adopting a scheme to subsidize the costs of hiring R&D personnel in SMEs. Such a scheme has been operating for seven or eight years in West Germany with, I believe, very considerable success. The scheme would help not just high-technology SMEs, but it would also make a contribution towards the technological upgrading of SMEs in general.

The need for management education also emerges clearly from the cases studied for IRDAC. In particular, a significant number of the sample firms expressed a need for education in the area of *technology management*. There seems little doubt that the most important factor influencing successful innovation and growth is the quality of management and EC action in this area could, I believe, be highly beneficial.

Finally, I would like to reinforce existing demands for the establishment of comprehensive and easily accessible data bases for SMEs. The data bases would need to contain a wide range of *technical, business and commercial* information.

Summary of Policy Recommendations

(a) Establish a "common" market with common standards and regulations;

(b) Adopt a SBIR-type scheme linked to a "hands-on" development capital fund;

(c) Facilitate SMEs' access to public procurement contracts;

(d) Facilitate SME interfirm joint R&D ventures. Develop special EC SME collaborative technology programmes;

(e) Subsidise the employment costs of R&D personnel in SMEs;

(f) Initiate (technology) management education programmes for SME managers;

(g) Establish comprehensive and easily accessible data bases for SMEs containing a range of technical, business and commercial information; and

(h) Establish an agency devoted to the wellbeing of SMEs, perhaps a European SBA.

The last recommendation (h) is an extremely important one. It is clear that what is required is not just the initiation of one or several measures by the EC, but rather an interrelated "package" of measures designed to accommodate the complex requirements of SME innovation and growth. A European SBA – or Agency for the Support of Small Business Expansion, or SME EIRMA, or whatever – could play a necessary *coordinating,* as well as lobbying and support, role thereby adding *coherence* to what might otherwise be a largely disparate set of initiatives. A properly constructed, coherent package of measures could play a crucial role in stimulating SME innovation and growth to help carry Europe successfully and competitively into the 21st Century.

Appendix 1

Trends in the Size Distribution of Innovating Firms (%) 1945 - 83

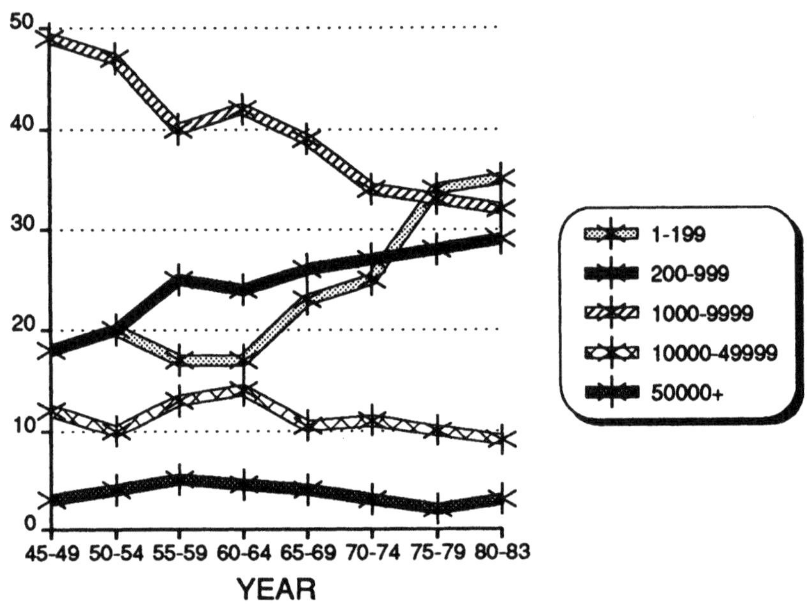

Appendix 2

NTBFs in Europe 1970-1985

	UK	FRG
Number in Existence:	7,000	3,000
Start-up Rate:	Explosive since the mid-1970s. Most are "very young" - post 1979	Mid 1970s: 40-60 per annum. 1983: 125
Employment. (1985)	About 120,000 Electronics; Computer h/w & s/w; instruments; (recently) biotechnology	About 120,000 Electrs & telecoms,50% Comp h/w & s/w,17% Sci instruments,16%

Strong geographical clustering effects in both countries
(Segal Quince and ISI Institute Study, 1986)

Appendix 3

NTBFs in the UK and the FRG, 1950-1975

	UK	FRG
Number in Existence:	200	100
	(considered only manufacturing – probably underestimated)	
Total Employment:	15,000	12,000
Total Annual Sales (1985)	200 million pounds (50% from 4 companies)	180 million pounds (60% from one company – Nixdorf)
Export Ratio:	30%	

Sectors: Mainly electrical and electronic

(A. D. Little, study, 1977)

Appendix 4

Emerging Technologies

Biotechnology	Single Cell Protein Bio-Engineering Biomass Diagnostic Kits Pharmaceuticals
Energy Technologies	Heat Pumps Solar Energy Devices Coal Gasification and Liquefaction Renewable Energy Sources Monitoring and Control Equipment
Advanced Materials Technology	Biocompatible Materials (Implants) Advanced Composite Material Advanced Electronics Materials Superconducting Materials
Information Technologies	Electronic Office Equipment Fibre Optic Systems Satellite Communications Scientific and Medical Instruments Advanced Computing Software Developments I.T. in the Home Etc. (I.T. for existing and new applications)

Appendix 5

Factors Encouraging NTBF Formation in USA

- Large and innovating home market (strong pull)
- Culture favourable to entrepreneurship (urge to succeed; low social penalty for failure)
- Mobility of key individuals: Job-hopping and many spin-offs
- Strong belief in the importance of small, independent companies: enshrined in the Small Business Act of 1953
- High availability of venture capital
- R&D support from DoD (and, more recently SBIR scheme)
- High level of public procurement
- High level if public support (DoD, NASA) for advanced technology in relation to position as a super power
- Large/small firm complementarities (spin-offs: purchasing; venture capital; joint ventures)
- Relatively strong university/industry linkages: entrepreneurial tendencies of academics

Appendix 6

Estimated Size of National Venture Capital Industries (1985/86)

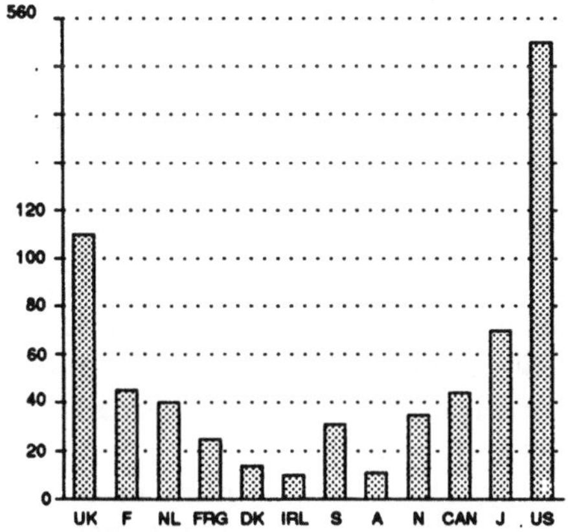

Number of Venture Capital firms

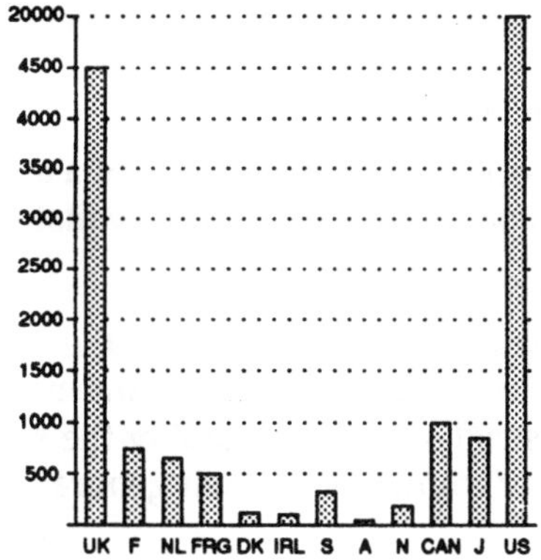

Total Venture Capital Pool (M$)

- EC total in 1985, $4.75bn: UK about 40% of total
- European VC growth rate in 1985, about 38%
- US is about 3 x European total – but , US VC is hands-on, with very considerable accumulated evaluation and management expertise and efficient inter-fund informal communications network

Towards a European Research and Policy Agenda for Small Countries and Economies

D. O'Doherty

The small countries of Europe have a diversity of strategy, structure and experience equivalent to that of the range of small firms which are such a large part of their economies. Some small countries are at the advanced industrialized end of the development spectrum while others have many of the characteristics of the developing countries. The former often have a number of large firms which are global "players" and which also have major impacts on local SMEs and on their local economies generally. The latter countries are normally dominated by very small firms, often in the more traditional sectors and industries.

Small countries do, however, have a number of significant similarities particularly in a "global" setting, which make it important to study their performance as a group, while being acutely conscious of their many dissimilarities and always seeing the need to also look at individual countries and at sub groups within the overall "small country" designation.

These common characteristics include

- a numerically small population, leading inter alia to a limited domestic market and a small amount of resources available for R&D;
- a high coefficient of external dependence;
- certain "peripheral area" features, marked in the case of some small countries by difficulties of access to important markets, as a consequence of their industrial structures notably the size of their firms, and in the case of other countries, by their geographical situation.

Small Country Characteristics

Structure

In addition to their obvious scale characteristics, especially regarding the small relative and absolute size of both firms and markets, small countries have limited manpower and technological resources. They suffer problems of "structural competitiveness" whereby their truncated industrial systems give them only limited scope for vertical integration and interaction between firms and sectors. They tend to specialise in particular product areas – allowing them an element of flexibility and responsiveness but also making them relatively vulnerable to technological and market changes. Some also have high levels of foreign investment penetration with its consequent advantages for international access to technologies and markets, but often meaning a high level of dependency and low retained value added.

Strategy

Of their nature, small countries are export-oriented and are increasingly open to the concept, if not the practice, of cooperation at both inter-firm and inter-country levels. Many have chosen the same strategic market segments/technologies as both large and other small countries but with "niche" approaches have attempted to avoid head-on competition. As technology "followers" they tend to emphasise diffusion, although conscious of the need to maintain a place in the technology generation process. Education/human capital is usually a major focus of public investment given the relative lack of other assets and resources and in general the state plays an active role in the industrial and economic affairs of most small countries.

Culture

Whether in response to their given structural and strategic characteristics or as a major factor in actually defining these characteristics small countries have evolved models or systems of innovation and change of varying transparency. These have particularly been observed and analyzed in the Nordic countries, although even there much work remains on fully defining these specifically "national" approaches to the development and application of technology. These innovation systems or models are seen to reflect the distinctive cultures of the various countries, such as specific user-producer interactions and a preference for working in smaller units, which are perhaps more clearly defined or more easily observed in a small country context. The significance of these "national models of innovation" for both the individual country and for its partners within the EC must be a clear focus of attention in any analysis of the small countries and science and technology.

Studies

In the light of the characteristics outlined above what are the S&T-related issues and policy concerns which arise in relation to small European and particularly EC countries in the evolving era of internationalization and globalization?

It is proposed that a number of studies, largely based on existing research material at the member-state level, should be commissioned to explore these small country/S&T-related interfaces. These studies would be undertaken in a form which would allow them to be integrated into an overall EC level "Status Report" and/or reports on different categories of technology and/or of small countries, depending on their economic and industrial structures and their S&T status and performance.

Issues

What is the significance in the 1990s of the scale/structural make-up, strategic policies adopted by, and the innovation cultures which characterise small European countries?

Are there particular technologies which fit the scale/structural, strategic and cultural characteristics of small countries, e.g.

- software development/microelectronics applications
- AMT/FMS
- telecommunications
- biotechnology applications?

Are there particular industries/sectors or "applications" areas in which small countries could have S&T-related competitive advantages in a global context, e.g.

- social services
- producer services including "financial services" and "production systems"
- electronic instrumentation/telecoms equipment
- natural resource-based activities e.g. health foods?

Are small countries characterised by active/passive/integrated/non-integrated approaches or responses to the current and future technological challenges?

- role of local MNEs
- role of foreign MNEs
- role of publicly-owned firms
- role of small firms
- role of state agencies and labs/government initiatives
- role of universities/HEIs

- role of private intermediaries/facilitators

Is there significant scope for "cooperative" approaches?

- firm-firm alliances / partnerships including research collaboration, licensing and JVs
- clubs/consortia e.g. "CRAFT"-type initiatives
- "subcontracting" relationships
- state/firm links
- informal links/networks
- university-industry interaction
- inter-country cooperation including large-small and small country - small county links and cooperation with European countries (e.g. on the lines of the Nordic experience in telecoms systems and standards)

What is the role of EC S&T/Industrial Policy?

- Should there be explicit provision for small firms and/or small countries in the R&D "Framework Programme"?
- Should "SPRINT" and other programmes for diffusion and dissemination of existing and new technology be greatly strengthened in the interests of the smaller member-states?
- Should there be an overall technological/industrial policy for Community industry/small firms/small countries along with fuller integration of the relevant Commission Services (on the model of the Interservice Group on Rural Development?)
- Relevance or otherwise of the "subsidiarity" principle, limiting the roles of local, national and international authorities in a small country context?

Notes on Contributors

Section 1:

Dr Michael Pierce is Chief Executive of Mentec International Limited – a Dublin-based company with subsidiaries in both Ireland and Britain. A former lecturer at Trinity College, Dublin, he is Chairman of the Irish Entrepreneurs Association.

Mr Robert Cochran is Managing Director of Generics Limited which is engaged in software engineering for the international electronics industry. A former Irish National Delegate to the EC's ESPRIT Programme, he is currently Chairman of the Irish Computer Services Association.

Mr Jim Mountjoy, formerly Chief Executive of Baltimore Technologies Limited, Dublin has recently left to set up a similar company offering a range of services to the international telecommunications industry. He has previously been involved in the definition of national and international programmes in telecommunications research.

Dr Tom Sommerlatte is a Corporate Vice President of Arthur D. Little, Inc., and Managing Director of the firm's operations in Germany. He has 20 years consultancy experience at Arthur D. Little. Based on his background in chemical engineering, systems analysis, and business administration, he has worked with a large number of European and US companies assisting them in improving their international operating effectiveness and strategic management. His main interest is the linkage between strategic management, the effective and innovative deployment of techniques and the implementation of change.

Professor Paolo Mariti is Professor of Industrial Economics and Policy at the University of Pisa, Italy. He is author of *Non Competitive Relationships Between Firms in Modern Industrial Systems* (Italian) and has written and researched on cooperation agreements and industrial organizations.

Mr Keith Dickson, Dr Helen Lawton Smith and **Mr Stephen Smith** are all at Kingston Polytechnic, Surrey, England where they teach and research in the Business Innovation Research Unit. They have presented papers at a wide range of conferences on small firms and cooperation and their book *Research Collaboration in Industry* is to be published by Frances Pinter, London, during 1990.

Mr Edouard Bourgeois is Head of Department at ANVAR, the French technology transfer agency, where he has expanded the activities to include promoting mediation at regional and national levels.

Mr Vitor Corado Simoes is with the Technical University of Lisbon and is an adviser to the Secretary of State for European Affairs. Formerly with the Portuguese Foreign Investment Institute he is a member of the CIT which manages the EC's SPRINT Programme. He specialises in international investment and technological relations and was a member of the Programme Committee of the Dublin Conference on the Cooperation Phenomenon.

Professor Tom Allen, Professor D.J. Cogan, Dr. Brendan O'Sullivan and **Dermot O'Doherty** are all associated with the Science Policy Research Centre at University College, Dublin, of which Professor Cogan is Director. Professor Allen is also Director of the Management of Technology Centre at the Sloan School of Management MIT and is well-known for his work on "technology gatekeepers" and other aspects of the technology transfer process.

Section 2:

Mr Arturo Garcia Arroyo is Director of Technological Research (DGXII-C) at the Commission of the European Communities. From January 1983 until he joined the Commission, he was Deputy Director General of the Spanish Ministry of Education and Science, first with responsibility for Scientific Information and Documentation and then for Coordination and Promotion of Research. He is the author of more than fifty scientific and technical papers on heat transfer and of about twenty publications on science policy.

Mr Bruce Goodman is head of the European Communities VALUE Programme, in the Directorate for "Exploitation of Research and Technology Development, Innovation and Technology Transfer" at DGXIII in Luxembourg. This programme is concerned with the exploitation of Community research results and is linked to the SPRINT Programme and other technology transfer activities.

Mr Gerrt Heikens is currently involved in the Programme for Business Cooperation in DGXXIII at the Commission of the European Communities which he joined in 1987. Previously he has held a number of assignments with the Ministry of Foreign Affairs in the Netherlands including Secretary for Economic and Commercial Affairs, Royal Netherlands Embassy, Singapore, and at the Directorate of Economic Cooperation at the Hague.

Mr Joseph Mulcahy coordinates Structural Fund operations in DGXVI – Regional Policies at the Commission of the European Communities. He has also been involved in the planning, financing and technical aspects of the Community's common transport policy. Previously he was in the Directorate for Scientific and Technological Affairs at OECD and with the Department of the Environment in Ireland.

Professor Olaf Meyer is Head of the EUREKA Secretariat, Brussels, since July 1989. Physicist by education he served 25 years in Swedish industry as industrial researcher, manager, division head and vice president. From 1986 to 1989 he was managing director of the Swedish Industrial Development Fund. In parallel, from 1982 to 1986 he served as assistant professor at the Lulea University, subject: Rock Machine Techniques.

Section 3:

Professor Bengt-Åke Lundvall and **Mr Bent Dalum** are at the Institute of Production, Aalborg University Denmark. Professor Lundvall is co-editor (with Christopher Freeman) of *Small Countries facing the Technological Revolution* (1988). He is also a consultant to the OECD and participates in the Monitor Programme of the EC as has Bent Dalum. Bent Dalum is also co-author of "Small Open Economies in the World Market for Electronics – the Case of the Nordic Countries" in the above volume.

Mr Philip Murtagh is a partner in the Irish technical consultancy firm Norcontel which has carried out work for public and private organizations in Ireland under the EC regional telecommunications infrastructure programme, STAR, and the South East Regional Technology Programme (SERTEC). He also participated in the most recent review of the ESPRIT Programme on behalf of the EC, specialising in the role of small firms and small economies in Information Technology.

Mr Dermot O'Doherty (editor of this book) is a member of the National Science and Technology Policy Unit of Eolas – the Irish Science and Technology Agency – and an associate of the Science Policy Research Centre, University College, Dublin. He specialises in innovation and technology transfer, with particular emphasis on the roles of small firms and small countries. As well as being a member of the Steering Committee of the 6 Countries Programme on Innovation, he is the Irish National Delegate to the EC Monitor Programme and the OECD Committee for Scientific and Technological Policy (CSTP). He was the Chairperson of the Programme Committee for the conference on which this book is based.

Section 4:

Dr Andreas Goerdeler is at the Ministry for Research and Technology (BMFT) in Bonn where he specialises in the technological activities of small and medium-sized firms. He has been involved in the design and implementation of a number of measures to encourage small firm innovation and cooperation.

Mr Brian Arthur is with the Department of Trade and Industry (DTI) in London where he is in charge of long-range planning in the Research and Technology Policy Division. He is a member of the EC's Monitor Committee and current chairman of the 6 Countries Programme on Innovation.

Ms Isabel Matalonga, Dr R Guimaraes, Dr Mario de Abreu and **Ms A.P. Coelho** are all at the National Laboratory for Industrial and Engineering Technology (LNETI) in Lisbon, where they work on international collaboration and the planning of national and international technology activities. They are members of a number of international committees and are particularly interested in "small country" cooperation.

Mr Jaume Valls Passola is based at the Polytechnical University of Catalonia in Barcelona where he carries out a wide range of research on technological cooperation and regional development.

Mr Dermot O'Doherty – under Section 3 above.

Mr Liam Kelleher, Mr Tom McGuinness and **Austen Duke** are with the following Irish agencies – Córás Tráchtála (Export Board); Irish Productivity Centre, and Eolas (Science & Technology Agency) respectively. They offer a wide range of cooperation-related services to Irish industry.

Section 5:

Professor Roy Rothwell currently leads the Management of Technology Group at the Science Policy Research Unit of the University of Sussex. He joined the unit in 1971 to work on a comparative study of success and failure in industrial innovation. He has been a consultant to the OECD, EC and UNCTAD on various issues of technology and public policy and has written more than a hundred articles on the subject.

Mr Dermot O'Doherty – under Section 3 above.

Appendix 1

This appendix provides a selection of 'definitions' and 'typologies' of formal and informal arrangements, drawn from the literature on inter-firm cooperation.

Towards a working definition of cooperation agreements

Among industrial economists, one of the first and most careful general definitions is that of Mariti and Smiley (1983), for whose purposes, " a cooperative agreement is any long term, explicit agreement amongst tow or more firms. This agreement may or may not involve financial remuneration. There can be payment for some good or service: alternatively, the firms may agree to exchange information or some other commodity or service: alternatively, the firms may agree to exchange information or some other commodity or service: both are cooperative agreements. To meet this definition, the agreement must be long term must be long term: a one time purchase of goods and services is not a cooperative agreement, but an agreement to purchase all inputs from one supplier over the next ten years is a cooperative agreement (...). Cooperative agreements can be made verbally, although most agreements are indeed based on a written contract. Finally, a cooperative agreement can take various legal forms (...)". Since Mariti and Smiley's analysis is set against the classical distinction coined by Williamson (1975) between markets and hierarchies, they go on to specify that their objective is to study "the transactional middle ground between single transactions in the marketplace and internal organization which often occurs through merger, acquisitions and mergers are not treated as cooperative agreements".

This paper adopts the shorter, but rather similar definition developed by Ricotta and Mariotti and used by FOR (1986) to build their basic data base. The data base comprises "inter company cooperation agreements, which are formal, and informal, agreements between two or more companies providing for a certain degree of collaboration between them (and) involving equity participation or the creation of new companies (as well as) non-equity agreements". A wide range of interfirm cooperation agreements can be encompassed by this definition from which only outright acquisitions or mergers are excluded.

This exclusion is necessary for a number of reasons. Analytically (see integration and hierarchies in Williamson's definition and that of agreements. It is the only way of approaching the discussion regarding the novelty of agreements, in contrast with direct investments, acquisitions and mergers. It also represents a well established cut-off point within anti-trust and restrictive business practice theory and practice. In practice of course, the frontiers between the two may often be somewhat blurred, notably in the case of mergers which involve the takeover of strong technologically advanced corporations by conglomerates or very large corporations active in a different sector (e.g. the acquisition of Hughes Aerospace by General Motors). Likewise, within large industrial and financial conglomerate groups with loose corporate structures, agreements between firms and/or divisions may require almost as much negotiation as between totally independent, non-affiliated partners. Mergers may also prepare and/or follow agreements, or link up firms possessing complementary networks of agreements. As a result, in industry case studies it will often be necessary to include a consideration of mergers, but these must nonetheless be considered as conceptually different from agreements.

The distinction is particularly important when the international flow of technology and cooperation between firms of different nationalities are considered. Hierarchies, based on foreign direct investment, command intra-group scientific and technical cooperation between laboratories and firms which are situated in different countries but which belong to the same international corporate structure (see inter alia Michalet, 1976 and Caves, 1982). The same is true for international intra-group (intra-MNE) flows of technology. International interfirm cooperations call upon quite different mechanisms, be it between a large and a small firm or between two or more large firms, and also have distinctive policy implications.

License agreements lie at the other opposite side of the distinction between hierarchies and markets. In the area of technology, the outright acquisition of sale of a patent obviously falls quite definitely in the realm of the "market". Single one-time straightforward licensing agreements may also not be far from so doing. The case is already somewhat different when clauses providing for instance, for the communication by the licensee or by the licensor (or by both) of technological improvement, are included in contracts. Because of this the initiators of data bases on agreements almost invariably include licenses in their figures. A stricter count with two sets of numbers, including and excluding licenses, may be needed from now on.

Source: F. Chesnais, Technical Cooperation Agreements between Firms, *STI Review* No. 4, OECD 1988.

Many Partners	University based cooperation research is financed by firms (with or without public support)	R&D Cooperation	Precompetitive
	Government-industry cooperative R&D projects with universities and public research institute invlovement		
Several partners	Research and development corporations on a private joint venture basis		
Few or very few partners	Corporate venture capital in small high tech firms (By one or several firms otherwise competitors)	Technological Cooperation	Competitive
	Non-equity cooperative R&D agreements between two firms in selected areas		
	Technical agreements between firms concerning completed technology, inter alia: technologysharing agreements; second-sourcing agreements; complex two-way licensing; cross licensing in separate product markets		
Few or very partners	Industrial joint venture firms and comprehensive R&D manufacturing and marketing consortia	Manufacturing and/or marketing cooperation	
	Customer-supplier agreements, notably partnerships		
	One-way licensing and/or marketing agreements (incl. OEM sales agreements)		

Table 1: Interfirm research, technology and manufacturing cooperation agreements and the R&D, production and marketing spectrum

Source: F. Chesnais in *Technical Cooperation Agreements Between Firms: Some Initial Data and Analyses*, DSTI, 1986 on the basis of typologies by C. Hacklish (1986) and E. Ricotta and S. Mariotti (1986)

Types of Agreements

1. Technological cooperation: by this we mean the transfer of technology between companies, granting patents or the study of complementary technology.
2. Development cooperation: this concerns R&D agreements for developing new products.
3. Participation: this is taken as the acquisition of a minority equity interest.
4. Control: this is taken as the acquisition of the majority equity interest.
5. Incorporation: total take over of one company by another.
6. Merger: a new company is created out of the disappearance of pre-existing firms (or parts/departments of them).
7. Joint-venture: here it is considered as the creation of a new company or organization, while the contracting companies maintain their independence.
8. Commercial agreement: a traditional agreement for distributing and marketing products.

Source: R. Camagni, *Cooperation Agreements and New Forms for External Development of Companies*, IBM, 1988.

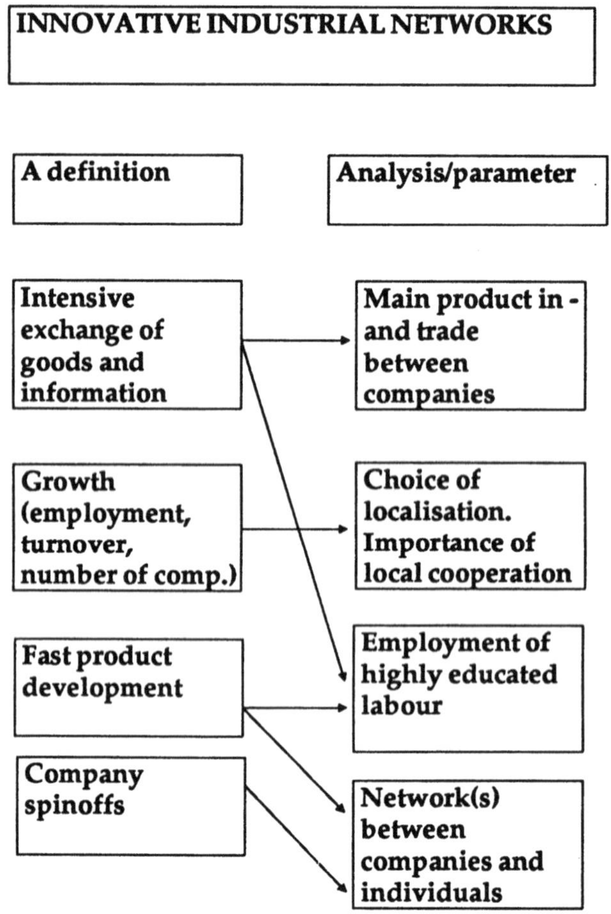

Figure 1: Innovative industrial networks
– Definitions and Parameters

Source: L. Gesling, "Knowledge Networks, Industrial Flexibility and Innovation" in F. Borum and P. H. Kristensen *Technological Innovation and Organizational Change* Danish Social Science Research Council 1989

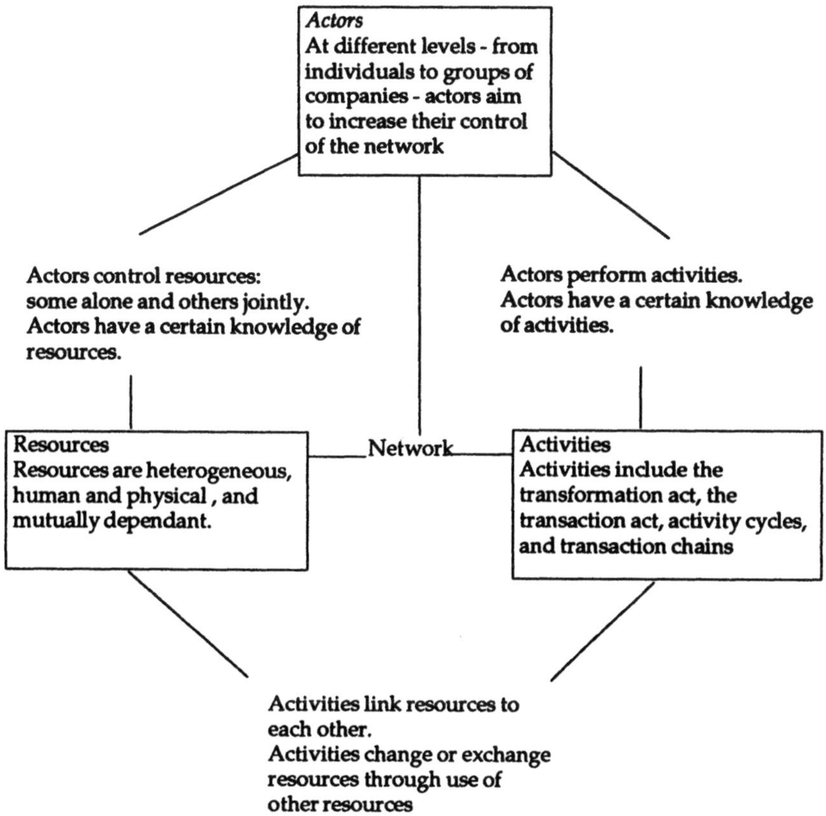

Figure 2: Network model

Source: H. Håkannson *Corporate Technological Behaviour – Cooperation and Networks* Routledge 1989

Appendix 2

This appendix outlines the different capabilities and involvements of small and large firms in innovation and cooperation activities.

	Small firms	Large firms
Marketing	Ability to react quickly to keep abreast of fast changing market requirements. (Market start-up abroad can be prohibitively costly.)	Comprehensive distribution and servicing facilities. High degree of market power with existing products.
Management	Lack of bureaucracy. Dynamic, entrepreneurial managers react quickly to take advantage of new opportunities and are willing to accept risk.	Professional managers able to control complex organizations and establish corporate strategies. (Can suffer an excess of bureaucracy. Ofter controlled by accountants who can be risk-averse. Managers can become mere 'administrators' who lack dynamism with respect to new long-term opportunities.)
Internal communications	Efficient and informal internal communication networks. Affords a fast response to internal problem-solving: provides ability to reorganize rapidly to adapt to change in the external environment.	(Internal communications often cumbersome: this can lead to slow reaction to external threats and opportunities.)
Qualified Technical Manpower	(Often lack suitable qualified technical specialists. Often unable to support a formal R&D effort on an appreciable scale.)	Ability to attract highly skilled technical specialists. Can support the establishment of a large R&D laboratory
External communications	(Often lack the time or resources to identify and use important external sources of scientific and technological expertise.)	Able to 'plug-in' to external sources of scientific and technological expertise. Can afford library and information services. Can subcontract R&D to specialist centres of expertise. Can buy crucial technical information and technology.
Finance	(Can experience great difficulty in attracting capital, especially risk capital. Innovation can represent a disproportionately large financial risk. Inability to spread risk over a portfolio of projects.)	Ability to borrow on capital market. Ability to spread risk over a portfolio of projects. Better able to fund diversification into new technologies and new markets.

Table 1: Advantages and disadvantages of small and large firms in innovation

	Small firms	Large firms
Economics of scale & systems approach	(In some areas scale economies form substantial entry barriers to small firms. Inability to offer integrated product lines or systems.)	Ability to gain scale economies in R&D production and marketing. Ability to offer a range of complementary products. Ability to bid for large turnkey projects.
Growth	(Can experience difficulty in acquiring external capital necessary for rapid growth. Entrepreneurial managers sometimes unable to cope with increasingly complex organizations.)	Ability to finance expansion of production base. Ability to fund growth via diversification and acquisition.
Patents	(Can experience problems in coping with the patent system. Cannot afford time or costs involved in patent litigation.)	Ability to employ patent specialists. Can afford to litigate to defend patents against infringement.
Government regulations	(Often cannot cope with complex regulations. Unit costs of compliance for small firms often high.)	Ability to fund legal services to cope with complex regulatory requirements. Can spread regulatory costs. Able to fund R&D necessary for compliances.

Table 1 (Contd): Advantages and disadvantages of small and large firms in innovation

Note: The statements in brackets represent areas of potential *disadvantage*. Abstracted from Rothwell and Zegveld (1982).

Source: R. Rothwell, Innovation and Firm Size: A Case for Dynamic Complementarity, *Journal of General Management*, Vol. 8, No.. 3. Spring 1983.

Manufacturing subcontracting relationships
SMFs supply components and sub-assembly to large companies. As part of this process large companies frequently transfer technological, manufacturing and quality control know-how to their small suppliers. Stable relationships can develop which are mutually advantageous.

Producer/customer relationships
SMFs supply finished products to large companies. Large companies can transfer technological know-how and supply suggestions for improvements to small suppliers based on user experience.

Licensing agreements
Large companies provide licences to small firms for innovative new developments. This frequently involves technology that the large company does not wish to exploit in-house but which it wishes to gain a financial return on. In some cases it can involve technology the large company will subsequently purchase in the form of equipment for in-house use, for example large process companies transferring new process control technology to small instrument companies.

Contract-out R&D
Large companies fund targeted R&D in small specialist consultancy companies, e.g. automobile companies funding R&D in specialist engine developers; pharmaceutical companies funding R&D in small biotechnology companies.

Collaborative development
Large and small companies collaborate in the development of a new product for the large company, e.g. small software or design houses collaborating respectively with large computer and automobile manufacturers.

Large/small firm joint ventures
Large and small firms collaborate in the development of an innovative new product containing technology new to the large partner. The large firm provides financial, manufacturing and marketing resources; the small firm provides specialist technological know-how and entrepreneurial dynamism. Generally the new products are complementary to the large firms' product range. They are manufactured by the small partners.

Educational acquisitions
Large companies acquire NTBFs to provide them with a window on new technology and an entree to new business areas. Examples of this are fairly common in the 'new-wave' biotechnology field.

Sponsored spin-outs
The large company offers financial backing for entrepreneurial employees to spin-out to form a new small firm to exploit technology developed within the parent company, but which is deemed unsuitable for in-house exploitation.

Table 2: Some modes of large/small firm interaction

Venture nurturing The large company offers not only financial support to the sponsored spin-out, but also access to managerial, marketing and manufacturing expertise and, if appropriate, to channels of distribution.
Independent spin-out assistance The large company offers technical assistance to an independent spin-out and sometimes acts as first customer for its products. Pre-payments can provide a crucial source of income to the new company.
Personnel secondment A number of large European companies have developed schemes to 'loan' experienced managers to assist new and existing SMFs in their locality.

Table 2 (Contd.): Some modes of large/small firm interaction

Source: R. Rothwell, "SMFs, Interfirm Relationships and Technological Change" Journal of Entrepreneurship and Regional Development (1) 1989 P.275-291

MIX
Papier aus verantwortungsvollen Quellen
Paper from responsible sources
FSC® C105338

If you have any concerns about our products,
you can contact us on
ProductSafety@springernature.com

In case Publisher is established outside the EU,
the EU authorized representative is:
Springer Nature Customer Service Center GmbH
Europaplatz 3, 69115 Heidelberg, Germany

Printed by Libri Plureos GmbH
in Hamburg, Germany